AC Electric Machines

Mehdi Rahmani-Andebili

AC Electric Machines

Practice Problems, Methods, and Solutions

 Springer

Mehdi Rahmani-Andebili
Department of Engineering and Physics
University of Central Oklahoma
Edmond, OK, USA

ISBN 978-3-031-15141-5   ISBN 978-3-031-15139-2   (eBook)
https://doi.org/10.1007/978-3-031-15139-2

This Springer imprint is published by the registered company Springer Nature Switzerland AG
The registered company address is: Gewerbestrasse 11, 6330 Cham, Switzerland

Preface

The course of AC electric machines is one of the fundamental courses of electric power engineering major taught for junior students. The subjects include the operation of transformers, induction machines, and synchronous machines.

Like the previously published textbooks, the textbook includes very detailed and multiple methods of problem solutions. It can be used as a practicing textbook by students and as a supplementary teaching source by instructors.

To help students study the textbook in the most efficient way, the exercises have been categorized in nine different levels. In this regard, for each problem of the textbook, a difficulty level (easy, normal, or hard) and a calculation amount (small, normal, or large) have been assigned. Moreover, in each chapter, problems have been ordered from the easiest problem with the smallest calculations to the most difficult problems with the largest calculations. Therefore, students are suggested to start studying the textbook from the easiest problems and continue practicing until they reach the normal and then the hardest ones. On the other hand, this classification can help instructors choose their desirable problems to conduct a quiz or a test. Moreover, the classification of computation amount can help students manage their time during future exams and instructors give the appropriate problems based on the exam duration.

Since the problems have very detailed solutions and some of them include multiple methods of solution, the textbook can be useful for the under-prepared students. In addition, the textbook is beneficial for knowledgeable students because it includes advanced exercises.

In the preparation of problem solutions, it has been tried to use typical methods to present the textbook as an instructor-recommended one. In other words, the heuristic methods of problem solution have never been used as the first method of problem solution. By considering this key point, the textbook will be in the direction of instructors' lectures, and the instructors will not see any untaught problem solutions in their students' answer sheets.

The Iranian University Entrance Exams for the master's and PhD degrees of electrical engineering major is the main reference of the textbook; however, all the problem solutions have been provided by me. The Iranian University Entrance Exam is one of the most competitive university entrance exams in the world that allows only 10% of the applicants to get into prestigious and tuition-free Iranian universities.

Edmond, OK, USA   Mehdi Rahmani-Andebili

The Other Works Published by the Author

The author has already published the books and textbooks below with Springer Nature.

Textbooks

DC Electric Machines, Electromechanical Energy Conversion Principles, and Magnetic Circuit Analysis- Practice Problems, Methods, and Solutions, *Springer Nature*, 2022.

Differential Equations- Practice Problems, Methods, and Solutions, *Springer Nature*, 2022.

Feedback Control Systems Analysis and Design- Practice Problems, Methods, and Solutions, *Springer Nature*, 2022.

Power System Analysis - Practice Problems, Methods, and Solutions, *Springer Nature*, 2022.

Advanced Electrical Circuit Analysis – Practice Problems, Methods, and Solutions, *Springer Nature*, 2022.

AC Electrical Circuit Analysis – Practice Problems, Methods, and Solutions, *Springer Nature*, 2021.

Calculus – Practice Problems, Methods, and Solutions, *Springer Nature*, 2021.

Precalculus – Practice Problems, Methods, and Solutions, *Springer Nature*, 2021.

DC Electrical Circuit Analysis – Practice Problems, Methods, and Solutions, *Springer Nature*, 2020.

Books

Applications of Artificial Intelligence in Planning and Operation of Smart Grid, *Springer Nature*, 2022.

Design, Control, and Operation of Microgrids in Smart Grids, *Springer Nature*, 2021.

Applications of Fuzzy Logic in Planning and Operation of Smart Grids, *Springer Nature*, 2021.

Operation of Smart Homes, *Springer Nature*, 2021.

Planning and Operation of Plug-in Electric Vehicles: Technical, Geographical, and Social Aspects, *Springer Nature*, 2019.

Contents

About the Author

Mehdi Rahmani-Andebili is an Assistant Professor in the Department of Engineering and Physics at University of Central Oklahoma, OK, USA. Before that, he was also an Assistant Professor in the Electrical Engineering Department at Montana Technological University, MT, USA, and the Engineering Technology Department at State University of New York, Buffalo State, NY, USA. He received his first M.Sc. and Ph.D. degrees in Electrical Engineering (Power System) from Tarbiat Modares University and Clemson University in 2011 and 2016, respectively, and his second M.Sc. degree in Physics and Astronomy from the University of Alabama in Huntsville in 2019. Moreover, he was a Postdoctoral Fellow at Sharif University of Technology during 2016–2017. As a professor, he has taught many courses and labs, including Power System Analysis, DC and AC Electric Machines, Feedback Control Systems Analysis and Design, Renewable Distributed Generation and Storage, Industrial Electronics, Analog Electronics, Electrical Circuits and Devices, AC Electrical Circuits Analysis, DC Electrical Circuits Analysis, Essentials of Electrical Engineering Technology, and Algebra- and Calculus-Based Physics. Dr. Rahmani-Andebili has more than 200 single-author and first-author publications, including journal papers, conference papers, textbooks, books, and book chapters. He is an IEEE Senior Member and the permanent reviewer of many credible journals. His research areas include Smart Grid, Power System Operation and Planning, Integration of Renewables and Energy Storages into Power System, Energy Scheduling and Demand-Side Management, Plug-in Electric Vehicles, Distributed Generation, and Advanced Optimization Techniques in Power System Studies.

Problems: Transformers

Abstract

In this chapter, the problems concerned with the single-phase and three-phase transformers and autotransformers are solved. The subjects include the determination of efficiency, maximum efficiency, all-day efficiency, different power losses, voltage regulation, load factor, and power factor of a transformer using its equivalent circuit at different load conditions. In this chapter, the problems are categorized in different levels based on their difficulty levels (easy, normal, and hard) and calculation amounts (small, normal, and large). Additionally, the problems are ordered from the easiest problems with the smallest computations to the most difficult problems with the largest calculations.

1.1. Determine the load factor (LF) of a transformer that results in its maximum efficiency, where its core and full-load copper power losses are about 1.8 kW and 3 kW, respectively.

   Difficulty level   ● Easy   ○ Normal   ○ Hard
   Calculation amount   ● Small   ○ Normal   ○ Large
   1) 77.5%
   2) 80%
   3) 100%
   4) 50%

1.2. A 1000/100 V transformer has the no-load current of 0.5 A at the power factor of 0.3. Determine the magnetizing component of no-load current.

   Difficulty level   ● Easy   ○ Normal   ○ Hard
   Calculation amount   ● Small   ○ Normal   ○ Large
   1) $0.15\ A$
   2) $0.353\ A$
   3) $0.477\ A$
   4) $0.5\ A$

1.3. Calculate the load factor (LF) of a transformer in which it has its maximum efficiency if the core and full-load copper power losses of transformer are about 1.6 kW and 2.5 kW, respectively.

   Difficulty level   ● Easy   ○ Normal   ○ Hard
   Calculation amount   ● Small   ○ Normal   ○ Large
   1) 80%
   2) 100%
   3) 75%
   4) 64%

© The Author(s), under exclusive license to Springer Nature Switzerland AG 2023
M. Rahimi-Andebili, *AC Electric Machines*, https://doi.org/10.1007/978-3-031-15139-2_1

1.4. The results of an open-circuit test on a single-phase, 10 kVA, 2200/220 V, 60 Hz transformer are as follows:

	O.C. test on low-voltage side
Voltage	220 V rating
Current	2.5 A
Power	100 W

Calculate the excitation current as a percentage of the rated current on the low-voltage side of transformer.

Difficulty level   ● Easy   ○ Normal   ○ Hard
Calculation amount   ● Small   ○ Normal   ○ Large

1) 55%
2) 11%
3) 22%
4) 5.5%

1.5. If the maximum efficiency of a transformer occurs at 90% load factor, determine the ratio of the core power loss to the copper power loss in this condition.

Difficulty level   ● Easy   ○ Normal   ○ Hard
Calculation amount   ● Small   ○ Normal   ○ Large

1) 0.81
2) 0.90
3) 0.70
4) 0.95

1.6. The equivalent series impedance of a transformer, based on its own rated values in per unit (p.u.) system, is $Z_{eq,old}^{pu}$. If the base values of transformer are doubled, calculate the new equivalent series impedance in per unit (p.u.).

Difficulty level   ● Easy   ○ Normal   ○ Hard
Calculation amount   ● Small   ○ Normal   ○ Large

1) $2Z_{eq,old}^{pu}$
2) $4Z_{eq,old}^{pu}$
3) $0.5Z_{eq,old}^{pu}$
4) $Z_{eq,old}^{pu}$

1.7. In a three-phase, 20 kV, 500 kVA transformer, the equivalent series resistance is about 1%. Determine the resistance in Ohm.

Difficulty level   ● Easy   ○ Normal   ○ Hard
Calculation amount   ○ Small   ● Normal   ○ Large

1) 8 Ω
2) 4 Ω
3) 16 Ω
4) 2 Ω

1.8. The results of open-circuit and short-circuit tests on a single-phase, 10 kVA, 2200/220 V, 60 Hz transformer are as follows:

	O.C. test on low-voltage side	S.C. test on high-voltage side
Voltage	220 V rating	150 V
Current	2.5 A	4.55 A rating
Power	100 W	215 W

Determine the power factor of transformer in the open-circuit and short-circuit tests, respectively.

Difficulty level   ○ Easy   ● Normal   ○ Hard
Calculation amount   ● Small   ○ Normal   ○ Large

1) 0.18 and 0.31
2) 0.36 and 0.62
3) 0.09 and 0.15
4) 0.36 and 0.31

1.9. A single-phase, 10 kVA, 2200/220 V, 60 Hz transformer is available. Determine the base voltage on the high-voltage side, base current on the low-voltage side, and base impedance on the high-voltage side.

Difficulty level   ○ Easy   ● Normal   ○ Hard
Calculation amount   ● Small   ○ Normal   ○ Large

1) 2200 V, 45.5 A, and 484 Ω
2) 220 V, 45.5 A, and 4.84 Ω
3) 2200 V, 4.55 A, and 484 Ω
4) 220 V, 4.55 A, and 4.84 Ω

1.10. A single-phase, 10 kVA, 2200/220 V, 60 Hz transformer has the exciting current of 0.25 A measured on the high-voltage side. Determine the exciting current in per unit (p.u.) value.

Difficulty level   ○ Easy   ● Normal   ○ Hard
Calculation amount   ● Small   ○ Normal   ○ Large

1) 1 $p.\,u.$
2) 0.0055 $p.\,u.$
3) 0.55 $p.\,u.$
4) 0.055 $p.\,u.$

1.11. A single-phase, 10 kVA, 2200/220 V, 60 Hz transformer has the equivalent series impedance of 10.4 + j31.3 Ω referred to the high-voltage side. Determine the equivalent series impedance in per unit (p.u.) value.

Difficulty level   ○ Easy   ● Normal   ○ Hard
Calculation amount   ● Small   ○ Normal   ○ Large

1) 0.0215 + j0.0647 $p.\,u.$
2) 0.104 + j0.313 $p.\,u.$
3) 10.4 + j31.3 $p.\,u.$
4) 0.215 + j0.647 $p.\,u.$

1.12. Determine the voltage regulation of a transformer at full-load condition and the power factor of 0.8 lagging that its equivalent series impedance is about 0.03 + j0.045 p. u.

Difficulty level   ○ Easy   ● Normal   ○ Hard
Calculation amount   ● Small   ○ Normal   ○ Large

1) 5.1%
2) −3.0%
3) −4.5%
4) −0.16%

1.13. In a 2200/220 V, 50 Hz transformer, the voltage of 220 V is applied on the high-voltage side of transformer to obtain the rated current for the short-circuit test. Calculate the maximum voltage regulation of transformer.

Difficulty level   ○ Easy   ● Normal   ○ Hard
Calculation amount   ● Small   ○ Normal   ○ Large

1) 5%
2) 1%
3) 2.2%
4) 10%

1.14. In a single-phase transformer, the ohmic and inductive voltage drops are about 3% and 5%, respectively. Calculate the voltage regulation of transformer at full-load condition and the power factor of 0.8 lagging.

Difficulty level   ○ Easy   ● Normal   ○ Hard
Calculation amount   ● Small   ○ Normal   ○ Large
1) 3.5%
2) 4.5%
3) 7.8%
4) 5.4%

1.15. In a transformer, for the 10% of rated voltage, the rated current is achieved in the short circuit test. The equivalent series resistance is about 5%. Calculate the voltage regulation of transformer at full-load condition and the power factor of 0.8 lagging.

Difficulty level   ○ Easy   ● Normal   ○ Hard
Calculation amount   ○ Small   ● Normal   ○ Large
1) 5%
2) 9.2%
3) 10%
4) 5.2%

1.16. A single-phase transformer has the equivalent series impedance of 5%. The voltage regulation of transformer at power factor of 0.8 leading is zero. Calculate the ohmic power loss at half-load condition.

Difficulty level   ○ Easy   ● Normal   ○ Hard
Calculation amount   ○ Small   ● Normal   ○ Large
1) 0.03 $p.\,u.$
2) 0.04 $p.\,u.$
3) 0.0075 $p.\,u.$
4) 0.01 $p.\,u.$

1.17. A single-phase, 10 kVA, 2200/220 V, 60 Hz transformer has the equivalent series impedance of $10.4 + j31.3\ \Omega$ referred to the high-voltage side. Determine its full-load copper power loss in per unit (p.u.) value.

Difficulty level   ○ Easy   ● Normal   ○ Hard
Calculation amount   ○ Small   ● Normal   ○ Large
1) 0.215 $p.\,u.$
2) 0.0215 $p.\,u.$
3) 2.15 $p.\,u.$
4) 0.00215 $p.\,u.$

1.18. The primary side of an ideal three-winding single-phase transformer has been connected to a 200 V power supply. Calculate the current of primary side if the secondary and tertiary sides are connected to a 2 kVA load with the power factor of 0.8 lagging and a 3 kVA load with the unity power factor, respectively.

Difficulty level   ○ Easy   ● Normal   ○ Hard
Calculation amount   ○ Small   ● Normal   ○ Large
1) 23.1 A
2) 23.8 A
3) 25.1 A
4) 27.5 A

1.19. In a single-phase 50 kVA transformer, the core power loss is about 1 kW and the maximum efficiency occurs at 70% load factor (LF). Calculate the efficiency of transformer for a purely resistive load at full-load condition.

Difficulty level   ○ Easy   ● Normal   ○ Hard
Calculation amount   ○ Small   ● Normal   ○ Large
1) 96.3%
2) 95.5%
3) 99.1%
4) 94.2%

1.20. If the maximum efficiency of a single-phase transformer with the equivalent series impedance of $0.02 + j0.02$ p. u. is 97%, determine the power factor (PF) and load factor (LF) of the load in this condition.
1) $PF = 1, LF = 100\%$
2) $PF = 1, LF = 75\%$
3) $PF = 1, LF = 77\%$
4) $PF = 0.7, LF = 79\%$

1.21. The results of an open-circuit test on a single-phase, 10 kVA, 2200/220 V, 60 Hz transformer are as follows:

	O.C. test on low-voltage side
Voltage	220 V rating
Current	2.5 A
Power	100 W

Calculate the core loss resistor and the magnetizing reactance referred to the high-voltage side of transformer.
1) 48.4 $k\Omega$, 8.94 $k\Omega$
2) 4.84 $k\Omega$, 0.894 $k\Omega$
3) 484 $k\Omega$, 89.4 $k\Omega$
4) 48.4 Ω, 8.94 Ω

1.22. The results of a short-circuit test on a single-phase, 10 kVA, 2200/220 V, 60 Hz transformer are as follows:

	S.C. test on high-voltage side
Voltage	150 V
Current	4.55 A rating
Power	215 W

Calculate the equivalent series impedance referred to the low-voltage side of transformer.
1) 0.104 $k\Omega$, 0.313 $k\Omega$
2) 0.104 Ω, 0.313 Ω
3) 1.04 Ω, 3.13 Ω
4) 10.4 Ω, 31.3 Ω

1.23. Calculate the efficiency of a single-phase, 5 kVA, 240/120 V transformer for the output apparent power of 2 kVA at rated voltage and power factor of 0.8. Herein, assume that the transformer has the core power loss of 100 W at rated voltage and the copper power loss of 120 W at rated current.
1) 88%
2) 98%
3) 90%
4) 93%

1.24. When the secondary side of a single-phase transformer is short-circuited and a voltage of 30 V is applied on its primary side, the primary current and power consumption are 20 A and 200 W, respectively. Determine the series equivalent reactance referred to the primary side.

1) $1.41 \, \Omega$
2) $1.5 \, \Omega$
3) $1.71 \, \Omega$
4) $2.3 \, \Omega$

1.25. The maximum efficiency of a transformer with the power factor of 0.8 leading is 90%. Calculate the corresponding load factor if the core power loss of transformer is about 2%.

Difficulty level   ○ Easy   ● Normal   ○ Hard
Calculation amount   ○ Small   ● Normal   ○ Large

1) 90%
2) 75%
3) 45%
4) 80%

1.26. If the efficiency of a transformer with the unity power factor at full-load and half-load conditions is 80%, determine the equivalent resistance of transformer in per unit (p.u.) system.

Difficulty level   ○ Easy   ● Normal   ○ Hard
Calculation amount   ○ Small   ● Normal   ○ Large

1) $\frac{1}{27} \, p.u.$
2) $\frac{2}{27} \, p.u.$
3) $\frac{1}{6} \, p.u.$
4) $\frac{1}{3} \, p.u.$

1.27. The maximum efficiency of a transformer at the power factor of 0.9 lagging is about 90%. If, in this condition, the copper power loss is about 2%, calculate the corresponding load factor.

Difficulty level   ○ Easy   ● Normal   ○ Hard
Calculation amount   ○ Small   ● Normal   ○ Large

1) 90%
2) 80%
3) 96%
4) 40%

1.28. In a single-phase 100 kVA transformer, the equivalent series impedance is about $0.01 + j0.04 \, p.\,u.$ The no-load power factor of transformer under the rated voltage and frequency is about 0.2. Moreover, the maximum efficiency of transformer occurs at full-load condition. Calculate the no-load current of transformer in per unit (p.u.) value.

Difficulty level   ○ Easy   ● Normal   ○ Hard
Calculation amount   ○ Small   ● Normal   ○ Large

1) $0.12 \, p.\,u.$
2) $0.05 \, p.\,u.$
3) $0.10 \, p.\,u.$
4) $0.03 \, p.\,u.$

1.29. Two single-phase transformers with the rated values of 100 kVA and 2% impedance and 150 kVA and 3% impedance are operated in parallel. Calculate the maximum power that can be drawn from them while none of them is overloaded.

Difficulty level   ○ Easy   ● Normal   ○ Hard
Calculation amount   ○ Small   ● Normal   ○ Large

1) 250 kVA
2) 300 kVA
3) 200 kVA
4) 400 kVA

1.30. How one per unit (p.u.) load can be divided between the two single-phase transformers with the rated values of 500 kVA and 4% impedance and 800 kVA and 3% impedance?

Difficulty level   ○ Easy   ● Normal   ○ Hard
Calculation amount   ○ Small   ● Normal   ○ Large

1) 68% and 32%
2) 32% and 68%
3) 57% and 43%
4) 43% and 57%

1.31. Two single-phase transformers with the rated values of 500 kVA and 4% impedance and 1000 kVA and 5% impedance are operated in parallel. Calculate the maximum allowable power that can be drawn from them.

Difficulty level   ○ Easy   ● Normal   ○ Hard
Calculation amount   ○ Small   ● Normal   ○ Large

1) 1300 kVA
2) 1500 kVA
3) 750 kVA
4) 1250 kVA

1.32. The equivalent series impedance referred to the high-voltage side of a transformer is $10.4 + j31.3\ \Omega$. In addition, the magnitude of rated voltage and rated current of transformer on the low-voltage side referred to the high-voltage side are 2200 V and 4.55 A, respectively. Calculate the voltage regulation of transformer when it supplies 75% of its rated load at the power factor of 0.6 lagging.

Difficulty level   ○ Easy   ○ Normal   ● Hard
Calculation amount   ○ Small   ● Normal   ○ Large

1) −4.86%
2) −2.43%
3) 4.86%
4) 2.43%

1.33. Solve problem 1.32 again but consider a leading power factor for the load.

Difficulty level   ○ Easy   ○ Normal   ● Hard
Calculation amount   ○ Small   ● Normal   ○ Large

1) 5.64%
2) −5.64%
3) −2.82%
4) 2.43%

1.34. The maximum efficiency of a transformer occurs at 80% load factor. If the core power loss of transformer increases about 20%, determine the new load factor in which the transformer has the maximum efficiency.

Difficulty level   ○ Easy   ○ Normal   ● Hard
Calculation amount   ○ Small   ● Normal   ○ Large

1) 78%
2) 98%
3) 88%
4) 85%

1.35. The maximum efficiency of a transformer that happens at unity power factor and rated load is 90%. Calculate the efficiency of transformer at the same power factor but at half-load condition.

Difficulty level   ○ Easy   ○ Normal   ● Hard
Calculation amount   ○ Small   ● Normal   ○ Large

1) 85.3%
2) 96.7%
3) 87.8%
4) 93.9%

1.36. The results of open-circuit and short-circuit tests of a 2200/220 V and 24 kVA transformer are as follows:

$$V_{oc} = 220 \ V, I_{oc} = 9.5 \ A, P_{oc} = 460.8 \ W$$

$$V_{sc} = 175 \ V, I_{sc} = 10.9 \ A, P_{sc} = 720 \ W$$

Calculate the amount of active power that must be drawn from the 220 V terminal of transformer so that its efficiency is maximum.
Difficulty level   ○ Easy   ○ Normal   ● Hard
Calculation amount   ○ Small   ● Normal   ○ Large
1) 19.2 kW
2) 22.78 kW
3) 24 kW
4) 28.8 kW

1.37. The equivalent resistance of a 2000/200 V, 20 kVA transformer is about 0.015 p.u., and its maximum efficiency occurs at the load current of 90 A on the secondary side. Calculate its efficiency at half-load condition and the power factor of 0.8.
Difficulty level   ○ Easy   ○ Normal   ● Hard
Calculation amount   ○ Small   ● Normal   ○ Large
1) 89.1%
2) 91.1%
3) 93.7%
4) 96.1%

1.38. In the short-circuit test of a 400/100 V, 5 kVA transformer, the power of 250 W is measured. If the equivalent series impedance referred to the high-voltage side is 3.2 Ω, calculate the power factor when the voltage regulation of transformer at full-load condition is zero.
Difficulty level   ○ Easy   ○ Normal   ● Hard
Calculation amount   ○ Small   ● Normal   ○ Large
1) 1
2) 0.866 lagging
3) 0.5 leading
4) 0.866 leading

1.39. In a single-phase transformer, the copper power loss at 60% of full-load condition is about 0.0108 p.u. In addition, the voltage regulation of transformer at full-load condition with the power factor of 0.8 lagging is about 4.08%. Calculate the load factor in which the maximum voltage regulation occurs.
Difficulty level   ○ Easy   ○ Normal   ● Hard
Calculation amount   ○ Small   ● Normal   ○ Large
1) 3%
2) 4%
3) 5%
4) 6%

1.40. The maximum efficiency of a single-phase 100 kVA transformer that occurs at 80% of full-load condition is about 90%. Herein, the load power factor is 0.9 and the short-circuit impedance of transformer is about 10%. Calculate the equivalent series reactance of transformer.
Difficulty level   ○ Easy   ○ Normal   ● Hard
Calculation amount   ○ Small   ● Normal   ○ Large
1) 6.25%
2) 7.8%
3) 8%
4) 10%

1.41. A 50 kVA, 2400/240 V transformer has the core power loss of 200 W at rated voltage and the copper loss of 500 W at full-load condition. In addition, its load cycle is as follows. Calculate its all-day efficiency.

Load factor	0%	50%	75%	100%	110%
Power factor	0	1	0.8 lag.	0.9 lag.	1
Duration (hour)	6	6	6	3	3

Difficulty level   ○ Easy   ○ Normal   ● Hard
Calculation amount   ○ Small   ○ Normal   ● Large
1) 88.88%
2) 92.65%
3) 98.35%
4) 99.10%

1.42. A two-winding transformer with the rated values of 60 kVA, 240/1200 V, 60 Hz is available. This transformer has the full-load efficiency of 96% and is supplying a load with the power factor of 0.8. Convert the two-winding transformer to an autotransformer with the voltage level of 1440/1200 V, and calculate its rated apparent power. In addition, what will be the efficiency of autotransformer for the same load?
Difficulty level   ○ Easy   ○ Normal   ● Hard
Calculation amount   ○ Small   ○ Normal   ● Large
1) 60 kVA, 99.3%
2) 60 kVA, 96%
3) 180 kVA, 96%
4) 360 kVA, 99.3 %

1.43. The maximum efficiency of a single-phase, 3.3/0.4 kV, 200 kVA transformer that happens at the load factor (LF) of 85% is 95%. Calculate the all-day efficiency of transformer if its load cycle is as follows. Herein, assume that the load is purely resistive.

No-load for about 6 hours
70% of full-load for about 6 hours
85% of full-load for about 6 hours
Full-load for about 6 hours

Difficulty level   ○ Easy   ○ Normal   ● Hard
Calculation amount   ○ Small   ○ Normal   ● Large
1) 95.6%
2) 95%
3) 94.1%
4) 93.4%

Solutions of Problems: Transformers

2

Abstract

In this chapter, the problems of the first chapter are fully solved, in detail, step-by-step, and with different methods.

2.1. Based on the information given in the problem, we have:

$$P_c = 1.8 \ kW \tag{1}$$

$$P_{cu,FL} = 3 \ kW \tag{2}$$

A transformer can be operated with its maximum efficiency if its load factor (LF) is equal to the following term:

$$LF_{\eta_{max}} = \sqrt{\frac{P_c}{P_{cu,FL}}} \tag{3}$$

Therefore, for this problem, we have:

$$LF_{\eta_{max}} = \sqrt{\frac{1.8}{3}} = 0.775 \Rightarrow LF_{\eta_{max}} = 77.5 \ \%$$

Choice (1) is the answer.

2.2. Based on the information given in the problem, we have:

$$I_{NL} = 0.5 \ A \tag{1}$$

$$PF_{NL} = 0.3 \tag{2}$$

As we know, the relation below exists between the excitation current (I_o), that is, no-load current (I_{NL}), and its magnetizing component (I_m):

$$I_m = I_{NL} \sin \varphi_{NL} = I_{NL} \sqrt{1 - (\cos \varphi_{NL})^2} = \sqrt{1 - (PF_{NL})^2} \tag{3}$$

$$\Rightarrow I_m = 0.5 \times \sqrt{1 - 0.3^2} \Rightarrow I_m = 0.477 \ A$$

Choice (3) is the answer.

M. Rahmani-Andebili (ed.), *AC Electric Machines*, https://doi.org/10.1007/978-3-031-15139-2_2

2.3. Based on the information given in the problem, we have:

$$P_c = 1.6\ kW \tag{1}$$

$$P_{cu,FL} = 2.5\ kW \tag{2}$$

A transformer can work with its maximum efficiency if its load factor (LF) is equal to the following term:

$$LF_{\eta_{max}} = \sqrt{\frac{P_c}{P_{cu,FL}}} \tag{3}$$

Therefore, for this problem, we have:

$$LF_{\eta_{max}} = \sqrt{\frac{1.6}{2.5}} = 0.8 \Rightarrow LF_{\eta_{max}} = 80\%$$

Choice (1) is the answer.

2.4. The results of an open-circuit test on the low-voltage side of a single-phase, 10 kVA, 2200/220 V, 60 Hz transformer are as follows:

	O.C. test on low-voltage side
Voltage	220 V rating
Current	2.5 A
Power	100 W

Based on the information given in the problem, we have:

$$S = 10\ kVA \tag{1}$$

$$I_o = I_{oc} = 2.5\ A \tag{2}$$

$$V_L = 220\ V \tag{3}$$

Figure 2.1 shows the equivalent electrical circuit of a transformer for an open-circuit test. The rated current on the low-voltage side of transformer can be calculated as follows:

$$I_{rated,L} = \frac{S}{V_L} = \frac{10 \times 10^3}{220} = 45.5\ A \tag{4}$$

The excitation current as a percentage of the rated current on the low-voltage side can be calculated as follows:

$$\frac{I_o}{I_{rated,L}} \times 100 = \frac{2.5}{45.5} \times 100 = 5.5\%$$

Choice (4) is the answer.

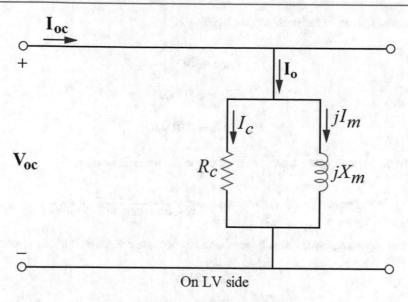

Fig. 2.1 The circuit of solution of Problem 2.4

2.5. Based on the information given in the problem, we have:

$$LF_{\eta_{max}} = 90\% \tag{1}$$

A transformer can work with its maximum efficiency if its load factor (LF) is equal to the following term:

$$LF_{\eta_{max}} = \sqrt{\frac{P_c}{P_{cu,FL}}} \tag{2}$$

Solving (1) and (2):

$$0.9 = \sqrt{\frac{P_c}{P_{cu,FL}}} \Rightarrow \frac{P_c}{P_{cu,FL}} = 0.81$$

Choice (1) is the answer.

2.6. Based on the information given in the problem, we have:

$$V_{B,new} = 2V_{B,old} \tag{1}$$

$$S_{B,new} = 2S_{B,old} \tag{2}$$

As we know, a per unit (p.u.) quantity can be updated based on the new base values as follows:

$$Z_{eq,new}^{pu} = \left(\frac{S_{B,new}}{S_{B,old}}\right)\left(\frac{V_{B,old}}{V_{B,new}}\right)^2 Z_{eq,old}^{pu} \tag{3}$$

Therefore:

$$Z_{eq,new}^{pu} = (2)\left(\frac{1}{2}\right)^2 Z_{eq,old}^{pu} \Rightarrow Z_{eq,new}^{pu} = 0.5 Z_{eq,old}^{pu}$$

Choice (3) is the answer.

2.7. Based on the information given in the problem, we have:

$$V_{rated} = 20 \; kV \tag{1}$$

$$S_{rated} = 500 \; kVA \tag{2}$$

$$R_{eq}^{pu} = 1\% = 0.01 \; p.u. \tag{3}$$

To present a quantity in per unit (p.u.) value, we need to use the formula below:

$$\text{Per unit (p.u.) value of quantity} = \frac{\text{Quantity}}{\text{Base value of quantity}} \tag{4}$$

As we know, the rated apparent power and voltage of a transformer are chosen as the base power and voltage. Therefore:

$$V_B = 20 \; kV \tag{5}$$

$$S_B = 500 \; kVA \tag{6}$$

The base impedance can be determined by using apparent power formula as follows:

$$S_B = \frac{V_B^2}{Z_B} \Rightarrow Z_B = \frac{V_B^2}{S_B} = \frac{(20 \; kV)^2}{500 \; kVA} \Rightarrow Z_B = 800 \; \Omega \tag{7}$$

Therefore, we can write:

$$R_{eq}^{pu} = \frac{R_{eq}}{Z_B} \Rightarrow 0.01 = \frac{R_{eq}}{800} \Rightarrow R_{eq} = 8 \; \Omega$$

Choice (1) is the answer.

2.8. The results of open-circuit and short-circuit tests of a single-phase, 10 kVA, 2200/220 V, 60 Hz transformer are as follows:

	O.C. test on low-voltage side	S.C. test on high-voltage side
Voltage	220 V rating	150 V
Current	2.5 A	4.55 A rating
Power	100 W	215 W

Based on the information given in the problem, we have:

$$V_{oc} = 220 \; V \tag{1}$$

$$I_{oc} = 2.5 \; A \tag{2}$$

$$P_{oc} = 100 \; W \tag{3}$$

$$V_{sc} = 150 \; V \tag{4}$$

$$I_{sc} = 4.55 \; A \tag{5}$$

$$P_{sc} = 215 \; W \tag{6}$$

The power factor of open-circuit test can be calculated as follows:

$$P_{oc} = V_{oc}I_{oc}\cos\varphi_{oc} \tag{7}$$

$$\Rightarrow PF_{oc} = \frac{P_{oc}}{V_{oc}I_{oc}} = \frac{100}{220 \times 2.5} \Rightarrow PF_{oc} = 0.18 \tag{8}$$

Likewise, the power factor of short-circuit test can be calculated as follows:

$$P_{sc} = V_{sc}I_{sc}\cos\varphi_{sc} \tag{9}$$

$$\Rightarrow PF_{sc} = \frac{P_{sc}}{V_{sc}I_{sc}} = \frac{215}{150 \times 4.55} \Rightarrow PF_{sc} = 0.31$$

Choice (1) is the answer.

2.9. Based on the information given in the problem, we have:

$$S = 10\,kVA \tag{1}$$

$$V = 2200/220\,V \tag{2}$$

The turns ratio of transformer can be calculated as follows:

$$a = \frac{N_1}{N_2} = \frac{V_1}{V_2} = \frac{2200}{220} = 10 \tag{3}$$

As we know, the rated voltage on each side of a transformer is chosen as the base voltage on the corresponding side. Therefore, the base voltage on the high-voltage and low-voltage sides of transformer are as follows:

$$V_{B,H} = 2200\,V \tag{4}$$

$$V_{B,L} = 220\,V \tag{5}$$

Moreover, the rated apparent power of a transformer is chosen as the base power for both sides. Hence:

$$S_B = 10\,kVA \tag{6}$$

The base current on the low-voltage side can be determined by using apparent power formula as follows:

$$S_B = V_{B,L}I_{B,L} \Rightarrow I_{B,L} = \frac{S_B}{V_{B,L}} = \frac{10^4}{220} \Rightarrow I_{B,L} = 45.5\,A$$

The base impedance on the high-voltage side can be calculated by using the other form of apparent power formula as follows:

$$S_B = \frac{V_{B,H}^2}{Z_{B,H}} \Rightarrow Z_{B,H} = \frac{V_{B,H}^2}{S_B} = \frac{2200^2}{10^4} \Rightarrow Z_{B,H} = 484\,\Omega$$

Choice (1) is the answer.

2.10. Based on the information given in the problem, we have:

$$S = 10 \ kVA \tag{1}$$

$$V = 2200/220 \ V \tag{2}$$

$$I_{o,H} = 0.25 \ A \tag{3}$$

To present a quantity in per unit (p.u.) value, we need to use the formula below.

$$\text{Per unit (p.u.) value of quantity} = \frac{\text{Quantity}}{\text{Base value of quantity}}$$

As we know, the rated apparent power of a transformer is chosen as the base power for both sides. In addition, the rated voltage on each side of a transformer is chosen as the base voltage on the corresponding side. Thus, the base voltage on the high-voltage side and the base power of transformer are as follows:

$$S_B = 10 \ kVA \tag{4}$$

$$V_{B,H} = 2200 \ V \tag{5}$$

To determine the base current on the high-voltage side, we can use the apparent power formula as follows:

$$S_B = V_{B,H} I_{B,H} \Rightarrow I_{B,H} = \frac{S_B}{V_{B,H}} = \frac{10^4}{2200} \Rightarrow I_{B,H} = 4.55 \ A$$

The exciting current in per unit (p.u.) value can be calculated as follows:

$$I_{o,H}^{pu} = \frac{I_{o,H}}{I_{B,H}} = \frac{0.25}{4.55} \Rightarrow I_{o,H}^{pu} = 0.055 \ p.u.$$

Choice (4) is the answer.

2.11. Based on the information given in the problem, we have:

$$S = 10 \ kVA \tag{1}$$

$$V = 2200/220 \ V \tag{2}$$

$$\mathbf{Z_{eq,H}} = 10.4 + j31.3 \ \Omega \tag{3}$$

To present a quantity in per unit (p.u.) value, we need to use the formula below:

$$\text{Per unit (p.u.) value of quantity} = \frac{\text{Quantity}}{\text{Base value of quantity}}$$

As we know, the rated apparent power of a transformer is chosen as the base power for both sides. In addition, the rated voltage on each side of a transformer is chosen as the base voltage on the corresponding side. Thus, the base voltage on the high-voltage side and the base power of transformer are as follows:

$$S_B = 10 \ kVA \tag{4}$$

$$V_{B,H} = 2200 \ V \tag{5}$$

The base impedance on the high-voltage side can be calculated by using the apparent power formula as follows:

$$S_B = \frac{V_{B,H}^2}{Z_{B,H}} \Rightarrow Z_{B,H} = \frac{V_{B,H}^2}{S_B} = \frac{2200^2}{10^4} \Rightarrow Z_{B,H} = 484 \ \Omega \tag{6}$$

The equivalent series impedance in per unit (p.u.) value can be calculated as follows:

$$\mathbf{Z}_{eq,H}^{pu} = \frac{\mathbf{Z}_{eq,H}}{Z_{B,H}} = \frac{10.4 + j31.3}{484} \Rightarrow \mathbf{Z}_{eq,H}^{pu} = 0.0215 + j0.0647 \ p.u.$$

Choice (1) is the answer.

2.12. Based on the information given in the problem, we have:

$$LF = 1 \tag{1}$$

$$PF = 0.8 \ \text{Lagging} \tag{2}$$

$$\mathbf{Z}_{eq}^{pu} = 0.03 + j0.045 \ p.u. \tag{3}$$

As we know, the voltage regulation of a transformer can be calculated as follows:

$$VR\% = LF \times \left(R_{eq}^{pu} \cos\varphi + X_{eq}^{pu} \sin\varphi \right) \times 100 \tag{4}$$

Therefore:

$$VR\% = 1 \times \left(0.03 \times 0.8 + 0.045 \times \sqrt{1 - 0.8^2} \right) \times 100 \Rightarrow VR\% = 5.1\%$$

Choice (1) is the answer.

2.13. Based on the information given in the problem, we have:

$$V = 2200/220 \ V \tag{1}$$

$$V_{sc} = 220 \ V \ \text{on high} - \text{voltage side} \tag{2}$$

$$I_{sc} = I_{rated} \tag{3}$$

The percent impedance of transformer can be calculated as follows:

$$U_k\% = \frac{V_{sc}}{V_B} = \frac{220}{2200} = 10\% \tag{4}$$

The maximum voltage regulation of a transformer can be calculated as follows:

$$VR_{max}\% = U_k\% = Z_{eq}\% \tag{5}$$

$$\Rightarrow VR\%_{max} = 10\%$$

Choice (4) is the answer.

2.14. Based on the information given in the problem, we have:

$$\Delta V_R^{pu} = 3\% = 0.03 \; p.u. \tag{1}$$

$$\Delta V_X^{pu} = 5\% = 0.05 \; p.u. \tag{2}$$

$$LF = 1 \tag{3}$$

$$PF = 0.8 \text{ Lagging} \tag{4}$$

As we know, the voltage regulation of a transformer can be calculated as follows:

$$VR\% = LF \times (\Delta V_R^{pu} \cos \varphi + \Delta V_X^{pu} \sin \varphi) \times 100 \tag{5}$$

Therefore:

$$VR\% = 1 \times \left(0.03 \times 0.8 + 0.05 \times \sqrt{1 - 0.8^2}\right) \times 100 \tag{6}$$

$$\Rightarrow VR\% \approx 5.4\%$$

Choice (4) is the answer.

2.15. Based on the information given in the problem, we have:

$$V_{sc} = 0.1 V_{rated} \tag{1}$$

$$I_{sc} = I_{rated} \tag{2}$$

$$R_{eq}^{pu} = 5\% = 0.05 \; p.u. \tag{3}$$

$$LF = 1 \tag{4}$$

$$PF = 0.8 \text{ Lagging} \tag{5}$$

Figure 2.2 shows the equivalent electrical circuit of a transformer for a short-circuit test. By applying Ohm's law for the whole branch, we have:

$$Z_{eq} = \frac{V_{sc}}{I_{sc}} = \frac{0.1 V_{rated}}{I_{rated}} \tag{6}$$

To present a quantity in per unit (p.u.) value, we need to use the formula below:

$$\text{Per unit (p.u.) value of quantity} = \frac{\text{Quantity}}{\text{Base value of quantity}} \tag{7}$$

The base impedance can be determined as follows:

$$Z_B = \frac{V_{rated}}{I_{rated}} \tag{8}$$

Solving (6) and (8):

$$Z_{eq} = 0.1 Z_B \tag{9}$$

$$\Rightarrow Z_{eq}^{pu} = \frac{Z_{eq}}{Z_B} = 0.1 \; p.u. \tag{10}$$

However, the value of Z_{eq}^{pu} can be calculated easier based on the definition of percent impedance of transformer ($U_k\%$). Since during the short-circuit test, rated current flows through the winding for the 10% of rated voltage, the percent impedance of transformer is 10%. Therefore:

$$Z_{eq}^{pu} = \frac{U_k\%}{100} = \frac{10}{100} = 0.1 \; p.u. \tag{11}$$

As we know, the relation between the magnitude of impedance (Z_{eq}), resistance (R_{eq}), and reactance (X_{eq}) is as follows:

$$\left(Z_{eq}^{pu}\right)^2 = \left(R_{eq}^{pu}\right)^2 + \left(X_{eq}^{pu}\right)^2 \tag{12}$$

$$\Rightarrow X_{eq}^{pu} = \sqrt{0.1^2 - 0.05^2} = 0.086 \; p.u. \tag{13}$$

As we know, the voltage regulation of a transformer can be calculated as follows:

$$VR\% = LF \times \left(R_{eq}^{pu} \cos\varphi + X_{eq}^{pu} \sin\varphi\right) \times 100 \tag{14}$$

Therefore:

$$VR\% = 1 \times \left(0.05 \times 0.8 + 0.086 \times \sqrt{1 - 0.8^2}\right) \times 100 \tag{15}$$

$$\Rightarrow VR\% \approx 9.2\%$$

Choice (2) is the answer.

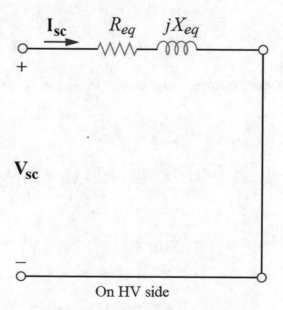

Fig. 2.2 The circuit of solution of Problem 2.15

2.16. Based on the information given in the problem, we have:

$$Z_{eq}\% = 5\% = 0.05 \, p.u. \tag{1}$$

$$VR\%_{PF=0.8 \text{ Leading}} = 0 \tag{2}$$

As we know, zero voltage regulation of transformer can occur when the load power factor is equal to the critical power factor (PF_{cr}) that can be calculated as follows:

$$PF_{cr} = \cos\varphi_{cr} = \frac{X_{eq}}{Z_{eq}} \tag{3}$$

In this condition, the load must have a leading power factor:
Therefore:

$$0.8 = \frac{X_{eq}}{0.05} \Rightarrow X_{eq} = 0.04 \, p.u. \tag{4}$$

As we know, the relation between the magnitude of impedance (Z_{eq}), resistance (R_{eq}), and reactance (X_{eq}) is as follows:

$$Z_{eq}^2 = R_{eq}^2 + X_{eq}^2 \tag{5}$$

$$\Rightarrow R_{eq} = \sqrt{0.05^2 - 0.04^2} = 0.03 \, p.u. \tag{6}$$

As we know, full-load copper power loss in per unit (p.u.) value is equal to equivalent series resistance in per unit (p.u.) value. In other words:

$$P_{cu,FL}^{pu} = R_{eq}^{pu} \tag{7}$$

Therefore:

$$P_{cu,FL}^{pu} = 0.03 \, p.u. \tag{8}$$

The copper power loss at full-load condition is now known. The copper power loss at half-load condition can be calculated as follows:

$$P_{cu,LF=50\%}^{pu} = LF^2 P_{cu,FL}^{pu} \tag{9}$$

$$\Rightarrow P_{cu,LF=50\%}^{pu} = (0.5)^2 \times 0.03 \Rightarrow P_{cu,LF=50\%}^{pu} = 0.0075 \, p.u.$$

Choice (3) is the answer.

2.17. Based on the information given in the problem, we have:

$$S = 10 \, kVA \tag{1}$$

$$V = 2200/220 \, V \tag{2}$$

$$Z_{eq,H} = 10.4 + j31.3 \, \Omega \tag{3}$$

First Method: The full-load copper power loss can be calculated as follows:

$$P_{cu,FL} = R_{eq,H} I_{FL,H}^2 \tag{4}$$

The current on the high-voltage side can be determined by using apparent power formula as follows:

$$S_{FL} = V_H I_{FL,H} \Rightarrow I_{FL,H} = \frac{S_{FL}}{V_H} = \frac{10^4}{2200} \Rightarrow I_{FL,H} = 4.55\ A \tag{5}$$

Solving (4) and (5):

$$P_{cu,FL} = 10.4 \times 4.55^2 = 215\ W \tag{6}$$

To present a quantity in per unit (p.u.) value, we need to use the formula below:

$$\text{Per unit (p.u.) value of quantity} = \frac{\text{Quantity}}{\text{Base value of quantity}} \tag{7}$$

As we know, the rated apparent power of a transformer is chosen as the base power for both sides. Therefore:

$$S_B = 10\ kVA \tag{8}$$

Hence:

$$P_{cu,FL}^{pu} = \frac{P_{cu,FL}}{S_B} = \frac{215}{10^4} \Rightarrow P_{cu,FL}^{pu} = 0.0215\ p.u. \tag{9}$$

Second Method: As we know, full-load copper power loss in per unit (p.u.) value is equal to equivalent series resistance in per unit (p.u.) value referred to either side. In other words:

$$P_{cu,FL}^{pu} = R_{eq,H}^{pu} = R_{eq,L}^{pu} \tag{10}$$

The base impedance on the high-voltage side can be calculated by using the apparent power formula as follows:

$$S_B = \frac{V_{B,H}^2}{Z_{B,H}} \Rightarrow Z_{B,H} = \frac{V_{B,H}^2}{S_B} = \frac{2200^2}{10^4} \Rightarrow Z_{B,H} = 484\ \Omega \tag{11}$$

$$R_{eq,H}^{pu} = \frac{R_{eq,H}}{Z_{B,H}} = \frac{10.4}{484} \Rightarrow R_{eq,H}^{pu} = 0.0215\ p.u.$$

Choice (2) is the answer.

2.18. Based on the information given in the problem, we have:

$$V_H = 200\ V \tag{1}$$

$$S_{L1} = 2\ kVA \tag{2}$$

$$PF_{L1} = 0.8\ \text{Lagging} \tag{3}$$

$$S_{L2} = 3 \, kVA \tag{4}$$

$$PF_{L2} = 1 \tag{5}$$

Since the transformer is ideal, it does not have any power loss. Hence, the input complex power is equal to the output complex power. In other words:

$$\mathbf{S_{input}} = \mathbf{S_{output}} \Rightarrow \mathbf{S_H} = \mathbf{S_{L1}} + \mathbf{S_{L2}} \Rightarrow \begin{cases} P_H = P_{L1} + P_{L2} \\ Q_H = Q_{L1} + Q_{L2} \end{cases} \tag{6}$$

For the active power, we have:

$$P_H = S_{L1}PF_{L1} + S_{L2}PF_{L2} = 2000 \times 0.8 + 3000 \times 1 \tag{7}$$

$$\Rightarrow P_H = 4600 \, W \tag{8}$$

For the reactive power, we have:

$$Q_H = S_{L1} \sin\left(\cos^{-1}PF_{L1}\right) + S_{L2} \sin\left(\cos^{-1}PF_{L2}\right) \tag{9}$$

$$\Rightarrow Q_H = 2000 \sin\left(\cos^{-1}0.8\right) + 3000 \sin\left(\cos^{-1}1\right) = 2000 \times 0.6 + 3000 \times 0 \tag{10}$$

$$\Rightarrow Q_H = 1200 \, W \tag{11}$$

The apparent power of transformer can be calculated as follows:

$$S_H = \sqrt{P_H^2 + Q_H^2} = \sqrt{4600^2 + 1200^2} = 4754 \, VA \tag{12}$$

The current can be determined by using apparent power formula as follows:

$$S_H = V_H I_H \Rightarrow I_H = \frac{S_H}{V_H} = \frac{4754}{200} \Rightarrow I_H = 23.8 \, A \tag{12}$$

Choice (2) is the answer.

2.19. Based on the information given in the problem, we have:

$$S = 50 \, kVA \tag{1}$$

$$P_c = 1 \, kW \tag{2}$$

$$LF_{\eta_{max}} = 70\% \tag{3}$$

A transformer can work with its maximum efficiency if its load factor (LF) is equal to the following term:

$$LF_{\eta_{max}} = \sqrt{\frac{P_c}{P_{cu,FL}}} \tag{4}$$

Therefore, for this problem, we have:

$$0.7 = \sqrt{\frac{1}{P_{cu,FL}}} \Rightarrow P_{cu,FL} \approx 2\ kW \tag{5}$$

The efficiency of a transformer, which is not operated at full load, can be determined as follows:

$$\eta = \frac{LF \times P_{out,FL}}{LF \times P_{out,FL} + P_c + LF^2 \times P_{cu,FL}} \times 100 = \frac{LF \times S_{out,FL} \times PF}{LF \times S_{out,FL} \times PF + P_c + LF^2 \times P_{cu,FL}} \times 100 \tag{6}$$

Therefore, the efficiency of transformer for a purely resistive load ($PF = 1$) and at full-load condition ($LF = 1$) can be calculated as follows:

$$\eta = \frac{1 \times 50 \times 1}{1 \times 50 \times 1 + 1 + 1^2 \times 2} \times 100 \Rightarrow \eta = 94.3\%$$

Choice (4) is the answer.

2.20. Based on the information given in the problem, we have:

$$\mathbf{Z_{eq}} = 0.02 + j0.02\ p.u. \tag{1}$$

$$\eta_{max} = 97\% \tag{2}$$

The maximum efficiency of a transformer occurs when the load is purely resistive and the load factor (LF) is equal to the specific load factor ($LF_{\eta_{max}}$). In other words:

$$PF = 1 \tag{3}$$

$$LF_{\eta_{max}} = \sqrt{\frac{P_c}{P_{cu,FL}}} \Rightarrow P_c = \left(LF_{\eta_{max}}\right)^2 \times P_{cu,FL} \tag{4}$$

Therefore, the maximum efficiency of a transformer, which is not operated at full load, can be determined as follows:

$$\eta_{max} = \frac{LF_{\eta_{max}} \times S_{out,FL} \times PF}{LF_{\eta_{max}} \times S_{out,FL} \times PF + P_c + \left(LF_{\eta_{max}}\right)^2 \times P_{cu,FL}} \times 100 \tag{5}$$

$$\xrightarrow{\times \frac{1}{S_{out,FL}}} \eta_{max} = \frac{LF_{\eta_{max}} \times PF}{LF_{\eta_{max}} \times PF + \frac{P_c}{S_{out,FL}} + \left(LF_{\eta_{max}}\right)^2 \times \frac{P_{cu,FL}}{S_{out,FL}}} \times 100 \tag{6}$$

As we know:

$$P_{cu,FL}^{p.u.} = \frac{P_{cu,FL}}{S_{out,FL}} \tag{7}$$

$$P_c^{p.u.} = \frac{P_c}{S_{out,FL}} \tag{8}$$

Solving (6)–(8):

$$\eta_{max} = \frac{LF_{\eta_{max}} \times PF}{LF_{\eta_{max}} \times PF + P_c^{p.u.} + \left(LF_{\eta_{max}}\right)^2 \times P_{cu,FL}^{p.u.}} \times 100 \tag{9}$$

Solving (4) and (9):

$$\eta_{max} = \frac{LF_{\eta_{max}} \times PF}{LF_{\eta_{max}} \times PF + 2 \times \left(LF_{\eta_{max}}\right)^2 \times P_{cu,FL}^{p.u.}} \times 100 \tag{10}$$

As we know, full-load copper power loss in per unit (p.u.) value is equal to equivalent series resistance in per unit (p.u.) value. In other words:

$$P_{cu,FL}^{pu} = R_{eq}^{pu} = 0.02 \, p.u. \tag{11}$$

Therefore:

$$0.97 = \frac{LF_{\eta_{max}} \times 1}{LF_{\eta_{max}} \times 1 + 2 \times \left(LF_{\eta_{max}}\right)^2 \times 0.02} \times 100 \Rightarrow LF_{\eta_{max}} = 0.77 = 77\%$$

Choice (3) is the answer.

2.21. The results of an open-circuit test on the low-voltage side of a single-phase, 10 kVA, 2200/220 V, 60 Hz transformer are as follows:

	O.C. test on low-voltage side
Voltage	220 V rating
Current	2.5 A
Power	100 W

Based on the information given in the problem, we have:

$$V_{oc} = 220 \, V \tag{1}$$

$$I_{oc} = 2.5 \, A \tag{2}$$

$$P_c = 100 \, W \tag{3}$$

Figure 2.3 shows the equivalent electrical circuit of a transformer for an open-circuit test. The relation between the core power loss and core loss resistance (R_c) is as follows.

$$P_c = \frac{(V_{oc})^2}{R_c} \tag{4}$$

$$\Rightarrow R_c = \frac{220^2}{100} = 484 \, \Omega \tag{5}$$

Applying Ohm's law for the core loss resistor:

$$I_c = \frac{V_{oc}}{R_c} \tag{6}$$

$$\Rightarrow I_c = \frac{220}{484} = 0.45 \, A \tag{7}$$

As we know, the relation between the magnitude of open-circuit current (I_{oc}), core loss current (I_c), and magnetizing current (I_m) is as follows:

$$I_{oc}^2 = I_c^2 + I_m^2 \tag{8}$$

$$I_m = \sqrt{I_{oc}^2 - I_c^2} \Rightarrow I_m = \sqrt{2.5^2 - 0.45^2} = 2.46 \, A \tag{9}$$

Applying Ohm's law for the magnetizing reactance:

$$X_m = \frac{V_{oc}}{I_m} = \frac{220}{2.446} = 89.4 \, \Omega \tag{10}$$

Since the open-circuit test has been done on the low-voltage side, the values of core loss resistance and magnetizing reactance are credible for the low-voltage side.

Referring the core loss resistance and magnetizing reactance to the high-voltage side:

$$R_{c,H} = R_{c,L} \times \left(\frac{V_H}{V_L}\right)^2 = 484 \times \left(\frac{2200}{220}\right)^2 \Rightarrow R_{c,H} = 48.4 \, k\Omega$$

$$X_{m,H} = X_{m,L} \times \left(\frac{V_H}{V_L}\right)^2 = 89.4 \times \left(\frac{2200}{220}\right)^2 \Rightarrow X_{m,H} = 8.94 \, k\Omega$$

Choice (1) is the answer.

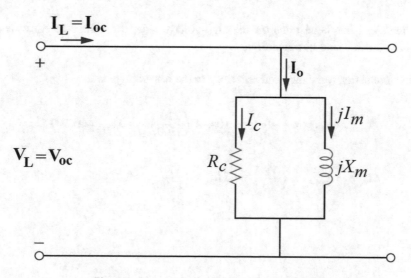

Fig. 2.3 The circuit of solution of Problem 2.21

2.22. The results of an short-circuit test on the high-voltage side of a single-phase, 10 kVA, 2200/220 V, 60 Hz transformer are as follows:

	S.C. test on high-voltage side
Voltage	150 V
Current	4.55 A rating
Power	215 W

Based on the information given in the problem, we have:

$$V_{sc} = 150 \ V \tag{1}$$

$$I_{sc} = 4.55 \ A \tag{2}$$

$$P_{cu} = 215 \ W \tag{3}$$

Figure 2.4 shows the equivalent electrical circuit of a transformer for a short-circuit test. The relation between the copper power loss and the equivalent series resistance (R_{eq}) is as follows:

$$P_{cu} = R_{eq}(I_{sc})^2 \tag{4}$$

$$\Rightarrow R_{eq} = \frac{215}{4.55^2} = 10.4 \ \Omega \tag{5}$$

Applying Ohm's law for the whole branch:

$$Z_{eq} = \frac{V_{sc}}{I_{sc}} = \frac{150}{4.55} = 32.97 \ \Omega \tag{6}$$

As we know, the relation between the magnitude of impedance (Z_{eq}), resistance (R_{eq}), and reactance (X_{eq}) is as follows:

$$Z_{eq}^2 = R_{eq}^2 + X_{eq}^2 \tag{7}$$

$$\Rightarrow X_{eq} = \sqrt{32.97^2 - 10.4^2} = 31.3 \ \Omega \tag{8}$$

Since the short-circuit test has been done on the high-voltage side, the values of equivalent series resistance and reactance are credible for this side of transformer.

Referring the equivalent series resistance and reactance to the low-voltage side:

$$R_{eq,L} = R_{eq,H} \times \left(\frac{V_L}{V_H}\right)^2 = 10.4 \times \left(\frac{220}{2200}\right)^2 \Rightarrow R_{eq,L} = 0.104 \ \Omega$$

$$X_{eq,L} = X_{eq,H} \times \left(\frac{V_L}{V_H}\right)^2 = 31.3 \times \left(\frac{220}{2200}\right)^2 \Rightarrow X_{eq,L} = 0.313 \ \Omega$$

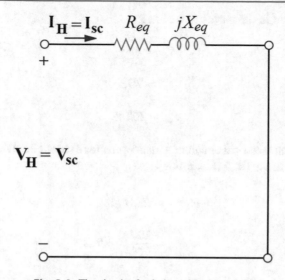

$$\mathbf{I_H = I_{sc}} \quad R_{eq} \quad jX_{eq}$$

$$\mathbf{V_H = V_{sc}}$$

Fig. 2.4 The circuit of solution of Problem 2.22

Choice (2) is the answer.

2.23. Based on the information given in the problem, we have:

$$S_{rated} = 5 \; kVA \tag{1}$$

$$S_{load} = 2 \; kVA \tag{2}$$

$$PF_{load} = 0.8 \tag{3}$$

$$P_{c,rated} = 100 \; W \tag{4}$$

$$P_{cu,rated} = 120 \; W \tag{5}$$

The efficiency of a transformer can be determined as follows:

$$\eta = \frac{P_{out}}{P_{out} + P_c + P_{cu}} \times 100 = \frac{S_{out} \cos\theta}{S_{out} \cos\theta + P_c + P_{cu}} \times 100 = \frac{S_{out} \times PF}{S_{out} \times PF + P_c + P_{cu}} \times 100 \tag{6}$$

Since the transformer is operated by its rated voltages, the core power loss will be constant. However, herein, the transformer is not operated by its rated load. Thus, its copper power loss needs to be updated as follows.

$$LF = \frac{S_{load}}{S_{rated}} = \frac{2}{5} = 0.4 \tag{7}$$

$$\Rightarrow P_{cu} = LF^2 P_{cu,rated} = 0.4^2 \times 120 \; W = 19.2 \; W \tag{8}$$

Therefore:

$$\eta = \frac{2 \times 10^3 \times 0.8}{2 \times 10^3 \times 0.8 + 100 + 19.2} \times 100 \Rightarrow \eta = 93\%$$

Choice (4) is the answer.

2.24. Based on the information given in the problem, we have:

$$V_{sc} = 30\ V \tag{1}$$

$$I_{sc} = 20\ A \tag{2}$$

$$P_{sc} = 200\ W \tag{3}$$

Figure 2.5 shows the equivalent electrical circuit of a transformer for a short-circuit test. The relation between the power and the equivalent series resistance (R_{eq}) is as follows.

$$P_{sc} = R_{eq}(I_{sc})^2 \tag{4}$$

$$\Rightarrow R_{eq} = \frac{200}{20^2} = 0.5\ \Omega \tag{5}$$

Applying Ohm's law for the whole branch:

$$Z_{eq} = \frac{V_{sc}}{I_{sc}} = \frac{30}{20} = 1.5\ \Omega \tag{6}$$

As we know, the relation between the magnitude of impedance (Z_{eq}), resistance (R_{eq}), and reactance (X_{eq}) is as follows:

$$Z_{eq}^2 = R_{eq}^2 + X_{eq}^2 \tag{7}$$

$$\Rightarrow X_{eq} = \sqrt{1.5^2 - 0.5^2} \Rightarrow X_{eq} = 1.41\ \Omega \tag{8}$$

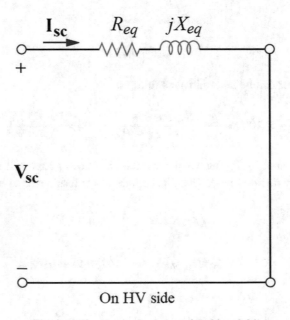

Fig. 2.5 The circuit of solution of Problem 2.24

Choice (1) is the answer.

2.25. Based on the information given in the problem, we have:

$$PF = 0.8 \text{ Leading} \tag{1}$$

$$\eta_{max} = 90\% \tag{2}$$

$$P_c^{pu} = 2\% = 0.02 \tag{3}$$

The maximum efficiency of a transformer, which is not operated at full load, can be determined as follows.

$$\eta_{max} = \frac{LF_{\eta_{max}} \times S_{out,FL} \times PF}{LF_{\eta_{max}} \times S_{out,FL} \times PF + 2P_c} \times 100 \tag{4}$$

$$\xrightarrow{\times \frac{1}{S_{out,FL}}} \eta_{max} = \frac{LF_{\eta_{max}} \times PF}{LF_{\eta_{max}} \times PF + 2\frac{P_c}{S_{out,FL}}} \times 100 = \frac{LF_{\eta_{max}} \times PF}{LF_{\eta_{max}} \times PF + 2P_c^{pu}} \times 100 \tag{5}$$

Therefore:

$$90 = \frac{LF_{\eta_{max}} \times 0.8}{LF_{\eta_{max}} \times 0.8 + 2 \times 0.02} \times 100 \Rightarrow LF_{\eta_{max}} = 0.45 = 45\% \tag{}$$

Choice (3) is the answer.

2.26. Based on the information given in the problem, we have:

$$PF = 1 \tag{1}$$

$$\eta|_{LF=1} = 80\% \tag{2}$$

$$\eta|_{LF=0.5} = 80\% \tag{3}$$

The efficiency of a transformer, which is not operated at full load, can be determined as follows:

$$\eta = \frac{LF \times S_{out,FL} \times PF}{LF \times S_{out,FL} \times PF + P_c + LF^2 \times P_{cu,FL}} \times 100 \tag{4}$$

Solving (1), (2), and (4):

$$80 = \frac{1 \times S_{out,FL} \times 1}{1 \times S_{out,FL} \times 1 + P_c + 1^2 \times P_{cu,FL}} \times 100 \Rightarrow 20S_{out,FL} = 80P_c + 80P_{cu,FL} \tag{5}$$

Solving (1), (2), and (4):

$$80 = \frac{0.5 \times S_{out,FL} \times 1}{0.5 \times S_{out,FL} \times 1 + P_c + 0.5^2 \times P_{cu,FL}} \times 100 \Rightarrow 10S_{out,FL} = 80P_c + 20P_{cu,FL} \tag{6}$$

Solving (5) and (6):

$$P_{cu,FL} = \frac{1}{6}S_{out,FL} \Rightarrow P_{cu,FL}^{pu} = \frac{P_{cu,FL}}{S_{out,FL}} = \frac{1}{6} \, p.u. \tag{7}$$

As we know, full-load copper power loss in per unit (p.u.) value is equal to equivalent series resistance in per unit (p.u.) value. Therefore:

$$R_{eq}^{pu} = \frac{1}{6} \ p.u.$$

Choice (3) is the answer.

2.27. Based on the information given in the problem, we have:

$$PF = 0.9 \ \text{Lagging} \tag{1}$$

$$\eta_{max} = 90\% \tag{2}$$

$$P_{cu,\eta_{max}}^{pu} = 2\% = 0.02 \tag{3}$$

The maximum efficiency of a transformer, which is not operated at full load, can be determined as follows:

$$\eta_{max} = \frac{LF_{\eta_{max}} \times S_{out,FL} \times PF}{LF_{\eta_{max}} \times S_{out,FL} \times PF + 2P_{cu,\eta_{max}}} \times 100 \tag{4}$$

$$\xrightarrow{\times \frac{1}{S_{out,FL}}} \eta_{max} = \frac{LF_{\eta_{max}} \times PF}{LF_{\eta_{max}} \times PF + 2\frac{P_{cu,\eta_{max}}}{S_{out,FL}}} \times 100 = \frac{LF_{\eta_{max}} \times PF}{LF_{\eta_{max}} \times PF + 2P_{cu,\eta_{max}}^{pu}} \times 100 \tag{5}$$

Therefore:

$$90 = \frac{LF_{\eta_{max}} \times 0.9}{LF_{\eta_{max}} \times 0.9 + 2 \times 0.02} \times 100 \Rightarrow LF_{\eta_{max}} = 0.4 = 40\%$$

Choice (4) is the answer.

2.28. Based on the information given in the problem, we have:

$$\mathbf{Z_{eq}} = 0.01 + j0.04 \ p.u. \tag{1}$$

$$PF_{NL} = \cos\varphi_{NL} = 0.2 \tag{2}$$

$$LF_{\eta_{max}} = 1 \tag{3}$$

As we know, full-load copper power loss in per unit (p.u.) value is equal to equivalent series resistance in per unit (p.u.) value. Therefore:

$$P_{cu,FL}^{pu} = 0.01 \ p.u. \tag{4}$$

On the other hand, as we know, a transformer can work with its maximum efficiency if its load factor (LF) is equal to the following term:

$$LF_{\eta_{max}} = \sqrt{\frac{P_c}{P_{cu,FL}}} \tag{5}$$

Solving (3)–(5):

$$1 = \sqrt{\frac{P_c}{0.01}} \Rightarrow P_c = 0.01 \; p.u. \tag{6}$$

The power that a transformer consumes at no-load condition is equal to its core power loss. In other words:

$$P_{NL} = P_c \tag{7}$$

where:

$$P_{NL} = V_{NL} I_{NL} \cos \varphi_{NL} \xrightarrow{\times \frac{1}{V_{FL} I_{FL}}} \frac{P_{NL}}{V_{FL} I_{FL}} = \frac{V_{NL}}{V_{FL}} \frac{I_{NL}}{I_{FL}} \cos \varphi_{NL} \Rightarrow P_{NL}^{pu} = V_{NL}^{pu} I_{NL}^{pu} \cos \varphi_{NL} \tag{8}$$

In addition, as we know, a rated voltage is applied in a no-load test. Hence, $V_{NL}^{pu} = 1 \; p.u.$

Therefore:

$$0.01 = 1 \times I_{NL}^{pu} \times 0.2 \Rightarrow I_{NL}^{pu} = 0.05 \; p.u.$$

Choice (2) is the answer.

2.29. Based on the information given in the problem, we have:

$$S_{rated,1} = 100 \; kVA \tag{1}$$

$$Z_{eq,1}\% = 2\% \Rightarrow Z_{eq,1}^{pu} = 0.02 \; p.u. \tag{2}$$

$$S_{rated,2} = 150 \; kVA \tag{3}$$

$$Z_{eq,2}\% = 3\% \Rightarrow Z_{eq,2}^{pu} = 0.03 \; p.u. \tag{4}$$

First, we need to determine the impedance of transformers based on a common power as follows:

$$S_{B,new} = 150 \; kVA \Rightarrow \begin{cases} Z_{eq,1,new}^{pu} = Z_{eq,1}^{pu} \times \frac{S_{B,new}}{S_{B,1}} = 0.02 \times \frac{150}{100} = 0.03 \; p.u. & (5) \\[2mm] Z_{eq,2,new}^{pu} = Z_{eq,2}^{pu} \times \frac{S_{B,new}}{S_{B,2}} = 0.03 \times \frac{150}{150} = 0.03 \; p.u. & (6) \end{cases}$$

Since the transformer with the lower apparent power is overloaded earlier, we need to adjust the total load of system based on the capacity of this transformer as follows:

$$S_{rated,1} = \frac{Z_{eq,2,new}^{pu}}{Z_{eq,1,new}^{pu} + Z_{eq,2,new}^{pu}} \times S_{load} \tag{7}$$

$$\Rightarrow 100 = \frac{0.03}{0.02 + 0.03} \times S_{load} \Rightarrow S_{load} = 200 \; kVA$$

Choice (3) is the answer.

2.30. Based on the information given in the problem, we have:

$$S_{load} = 1 \; p.u. \tag{1}$$

$$S_{rated,1} = 500 \; kVA \tag{2}$$

$$Z_{eq,1}\% = 4\% \Rightarrow Z_{eq,1}^{pu} = 0.04 \; p.u. \tag{3}$$

$$S_{rated,2} = 800 \; kVA \tag{4}$$

$$Z_{eq,2}\% = 3\% \Rightarrow Z_{eq,2}^{pu} = 0.03 \; p.u. \tag{5}$$

First, we need to determine the impedance of transformers based on a common power as follows:

$$S_{B,new} = 800 \; kVA \Rightarrow \begin{cases} Z_{eq,1,ew}^{pu} = Z_{eq,1}^{pu} \times \dfrac{S_{B,new}}{S_{B,1}} = 0.04 \times \dfrac{800}{500} = 0.064 \; p.u. & (6) \\[4mm] Z_{eq,2,ew}^{pu} = Z_{eq,2}^{pu} \times \dfrac{S_{B,new}}{S_{B,2}} = 0.03 \times \dfrac{800}{800} = 0.03 \; p.u. & (7) \end{cases}$$

The load is divided between the parallel transformers as follows:

$$S_1 = \frac{Z_{eq,2,new}^{pu}}{Z_{eq,1,new}^{pu} + Z_{eq,2,new}^{pu}} \times S_{load} \tag{8}$$

$$\Rightarrow S_1 = \frac{0.03}{0.064 + 0.03} \times 1 = 0.32 \; p.u. \Rightarrow S_1 = 32\%$$

$$S_2 = S_{load} - S_1 = 1 - 0.32 = 0.68 \; p.u. \Rightarrow S_2 = 68\%$$

Choice (2) is the answer.

2.31. Based on the information given in the problem, we have:

$$S_{rated,1} = 500 \; kVA \tag{1}$$

$$Z_{eq,1}\% = 4\% \Rightarrow Z_{eq,1}^{pu} = 0.04 \; p.u. \tag{2}$$

$$S_{rated,2} = 1000 \; kVA \tag{3}$$

$$Z_{eq,2}\% = 5\% \Rightarrow Z_{eq,2}^{pu} = 0.05 \; p.u. \tag{4}$$

First, we need to determine the impedance of transformers based on a common power as follows:

$$S_{B,new} = 1000 \; kVA \Rightarrow \begin{cases} Z_{eq,1,new}^{pu} = Z_{eq,1}^{pu} \times \dfrac{S_{B,new}}{S_{B,1}} = 0.04 \times \dfrac{1000}{500} = 0.08 \; p.u. & (5) \\[4mm] Z_{eq,2,new}^{pu} = Z_{eq,2}^{pu} \times \dfrac{S_{B,new}}{S_{B,2}} = 0.05 \times \dfrac{1000}{1000} = 0.05 \; p.u. & (6) \end{cases}$$

Since the transformer with the lower apparent power is overloaded earlier, we need to adjust the total load of system based on the capacity of this transformer as follows:

$$S_{rated,1} = \frac{Z_{eq,2,new}^{pu}}{Z_{eq,1,new}^{pu} + Z_{eq,2,new}^{pu}} \times S_{load} \tag{7}$$

$$\Rightarrow 500 = \frac{0.05}{0.08 + 0.05} \times S_{load} \Rightarrow S_{load} = 1300\ kVA$$

Choice (1) is the answer.

2.32. Based on the information given in the problem, we have:

$$\mathbf{Z_{eq,H}} = (10.4 + j31.3) \tag{1}$$

$$V_L' = 2200\ V \tag{2}$$

$$I_{L,rated}' = 4.55\ A \tag{3}$$

$$LF = 0.75 \tag{4}$$

$$PF = 0.6\ \text{Lagging} \tag{5}$$

The magnitude of current of load on the low-voltage side referred to the high-voltage side can be calculated as follows:

$$I_L' = LF \times I_{L,rated}' = 0.75 \times 4.55\ A = 3.41\ A \tag{6}$$

The phase angle of current on the low-voltage side for the load power factor of 0.6 lagging can be calculated as follows if the voltage on low-voltage side is assumed as the voltage reference:

$$\theta = -\cos^{-1}PF = -\cos^{-1}0.6 = -53.13° \tag{7}$$

Therefore:

$$\Rightarrow \mathbf{I_L'} = 3.41\angle -53.13°\ A \tag{8}$$

Based on definition, voltage regulation of a transformer can be calculated as follows, in which V_H and V_L' are the magnitude of voltage of transformer on the high-voltage side and the magnitude of voltage of transformer on the low-voltage side referred to the high-voltage side, respectively:

$$VR\% = \left(\frac{V_H}{V_L'} - 1\right) \times 100 \tag{9}$$

Figure 2.6 shows the approximate equivalent circuit of a transformer referred to the high-voltage side. Herein, $\mathbf{V_H}$ can be calculated by applying a KVL as follows:

$$\mathbf{V_H} = \mathbf{Z_{eq,H}I_H} + \mathbf{V_L'} = \mathbf{Z_{eq,H}I_L'} + \mathbf{V_L'} \tag{10}$$

$$\mathbf{V_H} = (10.4 + j31.3)(3.41\angle -53.13°) + 2200\angle 0° = 2306.94\angle 0.9°\ V \tag{11}$$

Therefore:

$$VR\% = \left(\frac{2306.94}{2200} - 1\right) \times 100 \Rightarrow VR\% = 4.86\%$$

As can be seen, the value of voltage regulation for a lagging load is positive.

Choice (3) is the answer.

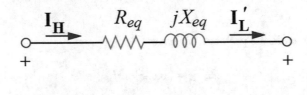

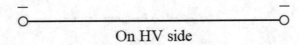

On HV side

Fig. 2.6 The circuit of solution of Problem 2.32

2.33. The phase angle of current on the low-voltage side for the load power factor of 0.6 leading can be calculated as follows if voltage on the low-voltage side is assumed as the voltage reference:

$$\theta = \cos^{-1} PF = \cos^{-1} 0.6 = 53.13° \tag{7}$$

Therefore:

$$\Rightarrow \mathbf{I'_L} = 3.41\angle 53.13\ A \tag{8}$$

Based on definition, voltage regulation of a transformer can be calculated as follows, in which V_H and V'_L are the magnitude of voltage of transformer on the high-voltage side and the magnitude of voltage of transformer on the low-voltage side referred to the high-voltage side, respectively:

$$VR\% = \left(\frac{V_H}{V'_L} - 1\right) \times 100 \tag{9}$$

Figure 2.7 shows the approximate equivalent circuit of a transformer referred to the high-voltage side. Herein, $\mathbf{V_H}$ can be calculated by applying a KVL as follows:

$$\mathbf{V_H} = \mathbf{Z_{eq,H} I_H} + \mathbf{V'_L} = \mathbf{Z_{eq,H} I'_L} + \mathbf{V'_L} \tag{10}$$

$$\mathbf{V_H} = (10.4 + j31.3)(3.41\angle 53.13) + 2200\angle 0 = 2137.9\angle 2.48° \tag{11}$$

Therefore:

$$VR\% = \left(\frac{2137.9}{2200} - 1\right) \times 100 \Rightarrow VR\% = -2.82\%$$

As can be seen, the value of voltage regulation for a leading load is negative.

Choice (3) is the answer.

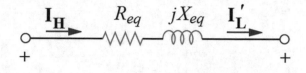

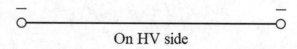

On HV side

Fig. 2.7 The circuit of solution of Problem 2.33

2.34. Based on the information given in the problem, we have:

$$LF_{\eta_{max}} = 80\% \tag{1}$$

$$P_{c,2} = 1.2 P_{c,1} \tag{2}$$

A transformer can work with its maximum efficiency if its load factor (LF) is equal to the following term:

$$LF_{\eta_{max}} = \sqrt{\frac{P_c}{P_{cu,FL}}} \tag{3}$$

For the first condition, we have:

$$0.8 = \sqrt{\frac{P_{c,1}}{P_{cu,FL}}} \tag{4}$$

For the second condition, we have:

$$LF_{\eta_{max},2} = \sqrt{\frac{P_{c,2}}{P_{cu,FL}}} \tag{5}$$

Solving (2) and (5):

$$LF_{\eta_{max},2} = \sqrt{\frac{1.2 P_{c,1}}{P_{cu,FL}}} = \sqrt{1.2}\sqrt{\frac{P_{c,1}}{P_{cu,FL}}} \tag{6}$$

Solving (4) and (6):

$$LF_{\eta_{max},2} = \sqrt{1.2} \times 0.8 = 0.88 \Rightarrow LF_{\eta_{max},2} = 88\%$$

Choice (3) is the answer.

2.35. Based on the information given in the problem, we have:

$$PF = 1 \tag{1}$$

$$LF_{\eta_{max}} = 1 \tag{2}$$

$$\eta_{max} = 90\% \tag{3}$$

A transformer can work with its maximum efficiency if its load factor (LF) is equal to the following term:

$$LF_{\eta_{max}} = \sqrt{\frac{P_c}{P_{cu,FL}}}$$

$$\Rightarrow 1 = \sqrt{\frac{P_c}{P_{cu,FL}}} \Rightarrow P_c = P_{cu,FL} \tag{4}$$

The maximum efficiency of a transformer, which is not operated at full load, can be determined as follows:

$$\eta_{max} = \frac{LF_{\eta_{max}} \times S_{out,FL} \times PF}{LF_{\eta_{max}} \times S_{out,FL} \times PF + 2P_c} \times 100 \tag{5}$$

Therefore:

$$\Rightarrow 90 = \frac{1 \times S_{out,FL} \times 1}{1 \times S_{out,FL} \times 1 + 2P_c} \times 100 \Rightarrow 0.9 = \frac{S_{out,FL}}{S_{out,FL} + 2P_c} \Rightarrow S_{out,FL} = 18P_c \tag{6}$$

The efficiency of a transformer, which is not operated at full load, can be determined as follows:

$$\eta = \frac{LF \times S_{out,FL} \times PF}{LF \times S_{out,FL} \times PF + P_c + LF^2 \times P_{cu,FL}} \times 100 \tag{7}$$

Calculating the efficiency of transformer at the same power factor but at half-load results:

$$\Rightarrow \eta = \frac{0.5 \times S_{out,FL} \times 1}{0.5 \times S_{out,FL} \times 1 + P_c + 0.5^2 \times P_{cu,FL}} \times 100 \tag{8}$$

Solving (4), (6), and (8):

$$\eta = \frac{0.5 \times 18P_c \times 1}{0.5 \times 18P_c \times 1 + P_c + 0.5^2 \times P_c} \times 100 = \frac{9}{10.25} \times 100 \Rightarrow \eta = 87.8\%$$

Choice (3) is the answer.

2.36. Based on the information given in the problem, the results of open-circuit and short-circuit tests of transformer are as follows:

$$V_{oc} = 220\ V, I_{oc} = 9.5\ A, P_{oc} = 460.8\ W \tag{1}$$

$$V_{sc} = 175\ V, I_{sc} = 10.9\ A, P_{sc} = 720\ W \tag{2}$$

Moreover, we have:

$$S_{FL} = 24\ kVA \tag{3}$$

The following terms can be extracted from the open-circuit and short-circuit tests of transformer:

$$P_c = 460.8\ W \tag{4}$$

$$P_{cu,FL} = 720\ W \tag{5}$$

As we know, a transformer can work with its maximum efficiency if its load factor (LF) is equal to the following term:

$$LF_{\eta_{max}} = \sqrt{\frac{P_c}{P_{cu,FL}}} \tag{6}$$

Therefore, we have:

$$LF_{\eta_{max}} = \sqrt{\frac{460.8}{720}} = 0.8 \tag{7}$$

Hence, the apparent power of transformer when it is operated with its maximum efficiency is as follows:

$$S_{\eta_{max}} = LF_{\eta_{max}} S_{FL} = 0.8 \times 24 = 19.2\ kVA \tag{8}$$

On the other hand, a resistive load can result in the maximum efficiency for the transformer. Therefore:

$$P_{\eta_{max}} = S_{\eta_{max}} PF_{\eta_{max}} = 19.2 \times 1 \Rightarrow P_{\eta_{max}} = 19.2\ kW$$

Choice (1) is the answer.

2.37. Based on the information given in the problem, we have:

$$V = 2000/200\ V \tag{1}$$

$$S_{FL} = 20\ kVA \tag{2}$$

$$R_{eq}^{pu} = 0.015\ p.u. \tag{3}$$

$$I_{2,\eta_{max}} = 90\ A \tag{4}$$

The rated current of transformer on the secondary side can be calculated as follows.

$$I_{2,FL} = \frac{S_{FL}}{V_2} = \frac{20000}{200} = 100\ A \tag{5}$$

The specific load factor, in which the transformer is operated with its maximum efficiency, can be calculated as follows:

$$LF_{\eta_{max}} = \frac{I_{2,\eta_{max}}}{I_{2,FL}} = \frac{90}{100} = 0.9 \tag{6}$$

As we know, a transformer can work with its maximum efficiency if its load factor (LF) is equal to the following term:

$$LF_{\eta_{max}} = \sqrt{\frac{P_c}{P_{cu,FL}}} \tag{7}$$

Solving (6) and (7):

$$\sqrt{\frac{P_c}{P_{cu,FL}}} = 0.9 \Rightarrow P_c = 0.81 P_{cu,FL} \tag{8}$$

On the other hand, as we know, full-load copper power loss in per unit (p.u.) value is equal to equivalent series resistance in per unit (p.u.) value. Therefore:

$$P_{cu,FL}^{pu} = 0.015 \, p.u. \tag{9}$$

$$P_{cu,FL} = P_{cu,FL}^{pu} \times S_{FL} = 0.015 \times 20 = 0.3 \, kW \tag{10}$$

Solving (8) and (10):

$$P_c = 0.81 P_{cu,FL} = 0.81 \times 0.3 = 0.243 \, kW \tag{11}$$

The efficiency of transformer in the new condition (at half-load condition and the power factor of 0.8) can be determined as follows:

$$\eta_{new} = \frac{LF_{new} \times S_{out,FL} \times PF_{new}}{LF_{new} \times S_{out,FL} \times PF_{new} + P_c + LF_{new}^2 \times P_{cu,FL}} \times 100 \tag{12}$$

$$\Rightarrow \eta_{new} = \frac{0.5 \times 20 \times 0.8}{0.5 \times 20 \times 0.8 + 0.243 + 0.5^2 \times 0.3} \times 100 \tag{13}$$

$$\Rightarrow \eta_{new} = 96.1\%$$

Choice (4) is the answer.

2.38. Based on the information given in the problem, we have:

$$V = 400/100 \, V \tag{1}$$

$$S_{rated} = 5000 \, kVA \tag{2}$$

$$P_{cu} = P_{sc} = 250 \, W \tag{3}$$

$$Z_{eq,H} = 3.2 \, \Omega \tag{4}$$

As we know, short-circuit test is done on the high-voltage side.
The rated current on the high-voltage side can be calculated as follows:

$$S_{rated} = V_{rated,H} I_{rated,H} \Rightarrow I_{rated,H} = \frac{S_{rated}}{V_{rated,H}} = \frac{5000}{400} = 12.5 \, A \tag{5}$$

Figure 2.8 shows the equivalent electrical circuit of a transformer for a short-circuit test. The relation between the copper power loss and the equivalent series resistance (R_{eq}) is as follows:

$$P_{cu} = R_{eq}(I_{sc})^2 \tag{6}$$

$$\Rightarrow R_{eq} = \frac{250}{12.5^2} = 1.6 \ \Omega \tag{7}$$

As we know, the relation between the magnitude of impedance (Z_{eq}), resistance (R_{eq}), and reactance (X_{eq}) is as follows:

$$Z_{eq}^2 = R_{eq}^2 + X_{eq}^2 \tag{8}$$

$$\Rightarrow X_{eq} = \sqrt{3.2^2 - 1.6^2} = 2.77 \ \Omega \tag{9}$$

As we know, zero voltage regulation of transformer can occur when the load power factor is equal to the critical power factor (PF_{cr}) that can be calculated as follows. In this condition, the load has a leading power factor:

$$PF_{cr} = \cos \varphi_{cr} = \frac{X_{eq}}{Z_{eq}} \tag{10}$$

$$\Rightarrow PF_{cr} = \frac{2.77}{3.2} \Rightarrow PF_{cr} = 0.866$$

Choice (4) is the answer.

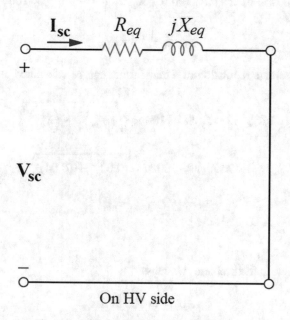

Fig. 2.8 The circuit of solution of Problem 2.38

2.39. Based on the information given in the problem, we have:

$$LF = 60\% \tag{1}$$

$$P_{cu,LF=60\%}^{pu} = 0.0108 \ p.u. \tag{2}$$

$$VR\%_{LF=100\%,PF=0.8 \ \text{Lag.}} = 4.08\% \tag{3}$$

The copper power loss at 60% of full-load condition has been given. The copper power loss at full-load condition can be calculated as follows:

$$P_{cu,LF=60\%}^{pu} = LF^2 P_{cu,FL}^{pu} \tag{4}$$

$$\Rightarrow 0.0108 = (0.6)^2 P_{cu,FL}^{pu} \Rightarrow P_{cu,FL}^{pu} = 0.03 \ p.u. \tag{5}$$

As we know, full-load copper power loss in per unit (p.u.) value is equal to equivalent series resistance in per unit (p.u.) value. In other words:

$$P_{cu,FL}^{pu} = R_{eq}^{pu} \tag{6}$$

Therefore:

$$R_{eq}^{pu} = 0.03 \ p.u. \tag{7}$$

As we know, the voltage regulation of a transformer can be calculated as follows:

$$VR\% = LF \times \left(R_{eq}^{pu} \cos \varphi + X_{eq}^{pu} \sin \varphi \right) \times 100 \tag{8}$$

Therefore:

$$4.08 = 1 \times \left(0.03 \times 0.8 + X_{eq}^{pu} \times \sqrt{1 - 0.8^2} \right) \times 100 \tag{9}$$

$$\Rightarrow X_{eq}^{pu} = 0.04 \ p.u. \tag{10}$$

As we know, the maximum voltage regulation of a transformer can be calculated as follows:

$$VR_{max}\% = Z_{eq}\% = 100 \times \sqrt{\left(R_{eq}^{pu} \right)^2 + \left(X_{eq}^{pu} \right)^2} \tag{11}$$

$$\Rightarrow VR\%_{max} = 100 \times \sqrt{(0.03)^2 + (0.04)^2} \tag{12}$$

$$\Rightarrow VR\%_{max} = 5\%$$

Choice (3) is the answer.

2.40. Based on the information given in the problem, we have:

$$S_{out,FL} = 100 \ kVA \tag{1}$$

$$LF_{\eta_{max}} = 0.8 \tag{2}$$

$$\eta_{max} = 90\% \tag{3}$$

$$PF = 0.9 \tag{4}$$

$$Z_{eq}\% = 10\% \tag{5}$$

The maximum efficiency of a transformer, which is not operated at full load, can be determined as follows:

$$\eta_{max} = \frac{LF_{\eta_{max}} \times S_{out,FL} \times PF}{LF_{\eta_{max}} \times S_{out,FL} \times PF + 2P_c} \times 100 \tag{6}$$

Therefore:

$$90 = \frac{0.8 \times 100 \times 0.9}{0.8 \times 100 \times 0.9 + 2P_c} \times 100 \Rightarrow P_c = 4\ kW \tag{7}$$

A transformer can work with its maximum efficiency if its load factor (LF) is equal to the following term:

$$LF_{\eta_{max}} = \sqrt{\frac{P_c}{P_{cu,FL}}} \tag{8}$$

Therefore, for this problem, we have:

$$0.8 = \sqrt{\frac{4}{P_{cu,FL}}} \Rightarrow P_{cu,FL} = 6.25\ kW \tag{9}$$

As we know, full-load copper power loss in per unit (p.u.) value is equal to equivalent series resistance in per unit (p.u.) value. In other words:

$$R_{eq}^{pu} = P_{cu,FL}^{pu} \tag{10}$$

where:

$$P_{cu,FL}^{pu} = \frac{P_{cu,FL}}{S_{out,FL}} = \frac{6.25\ kW}{100\ kVA} = 0.0625\ p.u. = 6.25\% \tag{11}$$

Therefore:

$$R_{eq}\% = 6.25\% \tag{12}$$

As we know, the relation between the magnitude of impedance (Z_{eq}), resistance (R_{eq}), and reactance (X_{eq}) is as follows:

$$Z_{eq}^2 = R_{eq}^2 + X_{eq}^2 \tag{13}$$

$$\Rightarrow X_{eq} = \sqrt{10^2 - 6.25^2} \Rightarrow X_{eq} = 7.8\%$$

Choice (2) is the answer.

2.41. Based on the information given in the problem, we have:

$$S_{FL} = 50\ kVA \tag{1}$$

$$P_c = 200\ W \tag{2}$$

$$P_{cu,FL} = 500\ W \tag{3}$$

Moreover, the load cycle of transformer is as follows:

Load factor	0%	50%	75%	100%	110%
Power factor	0	1	0.8 lag.	0.9 lag.	1
Duration (hour)	6	6	6	3	3

The all-day efficiency of a transformer can be determined as follows:

$$\eta_{AD} = \frac{E_{out}}{E_{in}} = \frac{\sum(P_{out,i}\Delta t_i)}{(P_{out,i}\Delta t_i) + \sum(P_{loss,i}\Delta t_i)} \tag{4}$$

$$\Rightarrow \eta_{AD} = \frac{\sum(S_{out,FL,i} \times LF_i \times PF_i \times \Delta t_i)}{\sum(S_{out,FL,i} \times LF_i \times PF_i \times \Delta t_i) + (P_c \times 24) + \sum(P_{cu,FL,i}LF_i^2\Delta t_i)} \tag{5}$$

where the description of parameters is as follows:

η_{AD}: All-day efficiency

E_{out}: Output energy (kWh)

E_{in}: Input energy (kWh)

P_{out}: Output power (kW)

Δt: Duration (hour)

P_{loss}: Power loss (kW)

$S_{out,\ FL}$: Full-load output apparent power (kVA)

LF: Load factor

PF: Power factor

P_c: Core power loss (kW)

$P_{cu,\ FL}$: Full-load copper loss (kW)

Therefore, we have:

$$\sum(S_{out,FL,i} \times LF_i \times PF_i \times \Delta t_i) = 50$$
$$\times 10^3((0 \times 0 \times 6) + (0.5 \times 1 \times 6) + (0.75 \times 0.8 \times 6) + (1 \times 0.9 \times 3) + (1.1 \times 1 \times 3))$$
$$= 630\ kWh$$

$$\tag{6}$$

$$P_c \times 24 = 200 \times 24 = 4.8\ kWh \tag{7}$$

$$\sum(P_{cu,FL,i}LF_i^2\Delta t_i) = 500((0^2 \times 6) + (0.5^2 \times 6) + (0.75^2 \times 6) + (1^2 \times 3) + (1.1^2 \times 3)) = 5.76\ kWh \tag{8}$$

$$\Rightarrow \eta_{AD} = \frac{630}{630 + 4.8 + 5.76} \times 100 \Rightarrow \eta_{AD} = 98.35\%$$

Choice (3) is the answer.

2.42. Based on the information given in the problem, we have:

$$S_{TW} = 60 \; kVA \tag{1}$$

$$V_{TW} = 240/1200 \; V \tag{2}$$

$$\eta_{TW} = 96\% \tag{3}$$

$$PF_{load} = 0.8 \tag{4}$$

$$V_{AT} = 1440/1200 \; V \tag{5}$$

The rated current of two-winding transformer on its primary and secondary sides can be calculated as follows:

$$I_{TW,1} = \frac{S_{TW}}{V_{TW,1}} = \frac{60 \times 10^3}{240} = 250 \; A \tag{6}$$

$$I_{TW,2} = \frac{S_{TW}}{V_{TW,2}} = \frac{60 \times 10^3}{1200} = 50 \; A \tag{7}$$

Figure 2.9 shows how we can covert the two-winding transformer with the voltage levels of 240/1200 V to an autotransformer with the voltage levels of 1440/1200 V. As can be seen in Fig. 2.9b, we have:

$$I_{AT,H} = I_{TW,1} = 250 \; A \tag{8}$$

$$V_{AT,H} = V_{TW,1} + V_{TW,2} = 240 + 1200 = 1440 \; V \tag{9}$$

Therefore, the rated apparent power of autotransformer can is as follows:

$$S_{AT} = V_{AT,H} I_{AT,H} = 1440 \times 250 = 360000 \; VA \Rightarrow S_{AT} = 360 \; kVA \tag{10}$$

The efficiency of a transformer at full-load condition can be calculated as follows:

$$\eta = \frac{P_{out}}{P_{out} + P_{loss}} \times 100 = \frac{S_{out} \cos\theta}{S_{out} \cos\theta + P_{loss}} \times 100 = \frac{S_{out} \times PF}{S_{out} \times PF + P_{loss}} \times 100 \tag{11}$$

Therefore, for the two-winding transformer, we can write:

$$96 = \frac{60 \times 0.8}{60 \times 0.8 + P_{TW,loss}} \times 100 \Rightarrow P_{TW,loss} = 2 \; kW \tag{12}$$

Since the frequency, voltages, and currents of windings are not changed, the core and copper losses will remain constant. In other words:

$$P_{AT,loss} = P_{TW,loss} = 2 \; kW \tag{13}$$

Now, the efficiency of autotransformer at full-load condition can be calculated as follows:

$$\eta_{AT} = \frac{S_{AT,out} \times PF}{S_{AT,out} \times PF + P_{AT,loss}} \times 100 = \frac{360 \times 0.8}{360 \times 0.8 + 2} \times 100 \tag{14}$$

$$\Rightarrow \eta_{AT} = 99.3\%$$

Choice (4) is the answer.

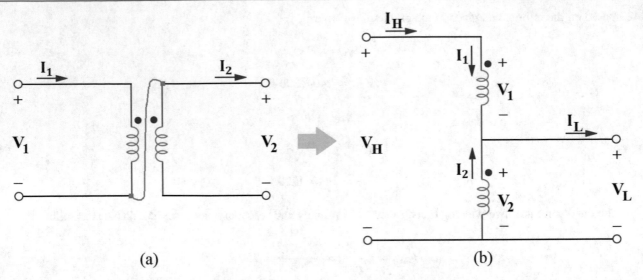

Fig. 2.9 The circuit of solution of Problem 2.42

2.43. Based on the information given in the problem, we have:

$$S = 200 \ kVA \tag{1}$$

$$LF_{\eta_{max}} = 0.85 \tag{2}$$

$$\eta_{max} = 95\% \tag{3}$$

$$PF = 1 \tag{4}$$

In addition, the load cycle of transformer is as follows:

Load factor (%)	0	70	85	100
Duration (hour)	6	6	6	6

The maximum efficiency of a transformer, which is not operated at full load, can be determined as follows:

$$\eta_{max} = \frac{LF_{\eta_{max}} \times S_{out,FL} \times PF}{LF_{\eta_{max}} \times S_{out,FL} \times PF + P_c + \left(LF_{\eta_{max}}\right)^2 \times P_{cu,FL}} \times 100 \tag{5}$$

Herein, $LF_{\eta_{max}}$ is the specific load factor that results in the maximum efficiency of transformer. In this condition, the relation below exists between the core power loss, full load copper loss, and $LF_{\eta_{max}}$:

$$LF_{\eta_{max}} = \sqrt{\frac{P_c}{P_{cu,FL}}} \Rightarrow P_c = \left(LF_{\eta_{max}}\right)^2 \times P_{cu,FL} \tag{6}$$

Solving (5) and (6):

$$\eta_{max} = \frac{LF_{\eta_{max}} \times S_{out,FL} \times PF}{LF_{\eta_{max}} \times S_{out,FL} \times PF + 2 \times \left(LF_{\eta_{max}}\right)^2 \times P_{cu,FL}} \times 100 \tag{7}$$

Therefore:

$$0.95 = \frac{0.85 \times 200 \times 1}{0.85 \times 200 \times 1 + 2 \times (0.85)^2 \times P_{cu,FL}} \times 100 \Rightarrow P_{cu,FL} = 6.2 \; kW \tag{8}$$

Solving (6) and (8):

$$P_c = (0.85)^2 \times 6.2 = 4.5 \; kW \tag{9}$$

The all-day efficiency of a transformer can be determined as follows:

$$\eta_{AD} = \frac{E_{out}}{E_{in}} = \frac{\sum(P_{out,i}\Delta t_i)}{(P_{out,i}\Delta t_i) + \sum(P_{loss,i}\Delta t_i)} \tag{10}$$

$$\Rightarrow \eta_{AD} = \frac{\sum(S_{out,FL,i} \times LF_i \times PF_i \times \Delta t_i)}{\sum(S_{out,FL,i} \times LF_i \times PF_i \times \Delta t_i) + (P_c \times 24) + \sum\left(P_{cu,FL,i}LF_i^2\Delta t_i\right)} \tag{11}$$

where the description of parameters is as follows:

η_{AD}: All-day efficiency

E_{out}: Output energy (kWh)

E_{in}: Input energy (kWh)

P_{out}: Output power (kW)

Δt: Duration (hour)

P_{loss}: Power loss (kW)

$S_{out, FL}$: Full-load output apparent power (kVA)

LF: Load factor

PF: Power factor

P_c: Core power loss (kW)

$P_{cu, FL}$: Full-load copper loss (kW)

Therefore, we have:

$$\sum(S_{out,FL,i} \times LF_i \times PF_i \times \Delta t_i)$$
$$= 200 \times 10^3((0 \times 1 \times 6) + (0.7 \times 1 \times 6) + (0.85 \times 1 \times 6) + (1 \times 1 \times 6)) \tag{12}$$
$$= 3060 \; kWh$$

$$P_c \times 24 = 4.5 \times 24 = 108 \; kWh \tag{13}$$

$$\sum\left(P_{cu,FL,i}LF_i^2\Delta t_i\right) = 6.2\left((0^2 \times 6) + (0.7^2 \times 6) + (0.85^2 \times 6) + (1^2 \times 6)\right) = 82.305 \; kWh \tag{14}$$

Therefore, the all-day efficiency of transformer can be calculated as follows:

$$\eta_{AD} = \frac{3060}{3060 + 108 + 82.305} \times 100 \Rightarrow \eta_{AD} = 94.1\%$$

Choice (3) is the answer.

Problems: Induction Machines

3

Abstract

In this chapter, the basic and advanced problems related to the three-phase wound rotor and squirrel-cage induction motors and generators are solved. The subjects include the calculation of slip, speed, electromotive force (emf), starting torque, maximum torque, rated torque, electromagnetic torque, mechanical torque, output torque, different powers and power losses, efficiency, maximum efficiency, and power factor of an induction machine using its equivalent circuit at different load and stability conditions. In this chapter, the problems are categorized in different levels based on their difficulty levels (easy, normal, and hard) and calculation amounts (small, normal, and large). Additionally, the problems are ordered from the easiest problems with the smallest computations to the most difficult problems with the largest calculations.

3.1. An induction motor with the star-connected stator is drawing 10 A current from the power grid. If its stator connection type is changed to delta, what will be the new current drawn from the grid?

Difficulty level   ● Easy   ○ Normal   ○ Hard
Calculation amount   ● Small   ○ Normal   ○ Large

1) $10\,A$
2) $10\sqrt{3}\,A$
3) $\frac{10}{\sqrt{3}}\,A$
4) $30\,A$

3.2. In a three-phase, six-pole, 400 V, 50 Hz induction motor, the frequency of electromotive force (emf) of rotor is about 100 cycle/minute. Calculate the slip and speed of rotor.

Difficulty level   ● Easy   ○ Normal   ○ Hard
Calculation amount   ○ Small   ● Normal   ○ Large

1) 0.021, $950\ rpm$
2) 0.021, $966\ rpm$
3) 0.033, $950\ rpm$
4) 0.033, $966\ rpm$

3.3. A 380 V induction motor has the starting torque equal to one and a half times of the rated torque. What voltage must be applied on the motor so that it develops the rated torque during the starting time?

Difficulty level   ○ Easy   ● Normal   ○ Hard
Calculation amount   ● Small   ○ Normal   ○ Large

1) $300\ V$
2) $310\ V$
3) $400\ V$
4) $380\ V$

3.4. In a three-phase, 220 *V* induction motor, the maximum torque is equal to three times of rated torque. Determine this ratio if the applied voltage is 110 *V*.

Difficulty level   ○ Easy   ● Normal   ○ Hard
Calculation amount   ● Small   ○ Normal   ○ Large

1) 8
2) 0.75
3) 4
4) 0.5

3.5. A three-phase, four-pole, 380 *V*, 50 Hz induction motor with a star-connected stator is being operated at rated load with the slip of 0.05. Calculate the speed of revolving fields of rotor and stator with respect to the rotor.

Difficulty level   ○ Easy   ● Normal   ○ Hard
Calculation amount   ○ Small   ● Normal   ○ Large

1) 75 *rpm*, 1500 *rpm*
2) 75 *rpm*, 0 *rpm*
3) 0 *rpm*, 75 *rpm*
4) 75 *rpm*, 75 *rpm*

3.6. A six-pole induction motor is supplied by a 60 Hz synchronous generator. If the motor is rotating with the speed of 1140 *rpm*, calculate the frequency of rotor current.

Difficulty level   ○ Easy   ● Normal   ○ Hard
Calculation amount   ○ Small   ● Normal   ○ Large

1) 2.4 *Hz*
2) 4 *Hz*
3) 1.8 *Hz*
4) 3 *Hz*

3.7. The input power, total power loss of stator, and rotational power loss of an induction motor are about 4 kW, 100 W, and 200 W, respectively. If the slip of machine is about 0.04, calculate its efficiency.

Difficulty level   ○ Easy   ● Normal   ○ Hard
Calculation amount   ○ Small   ● Normal   ○ Large

1) 89.4%
2) 88.6%
3) 90.2%
4) 85.5%

3.8. A three-phase, eight-pole, 400 *V*, 60 *Hz* induction motor is assumed. If the input power of rotor is about 90 kW, and the speed of revolving filed of stator with respect to the rotor is 10 rad/s, calculate the output power of motor. Herein, assume that the rotational power loss is ignorable.

Difficulty level   ○ Easy   ● Normal   ○ Hard
Calculation amount   ○ Small   ● Normal   ○ Large

1) 78.54 kW
2) 80.45 kW
3) 81.42 kW
4) 82.52 kW

3.9. A three-phase wound rotor induction motor has the stationary rotor impedance of $0.05 + j0.1$ *Ω/phase* referred to the stator side. Determine the external resistance of rotor so that the motor has the maximum torque during the starting time.

Difficulty level   ○ Easy   ● Normal   ○ Hard
Calculation amount   ○ Small   ● Normal   ○ Large

1) 0.075 Ω
2) 1.05 Ω
3) 0.1 Ω
4) 0.05 Ω

3.10. If the applied voltage of an induction motor decreases about 10%, how much decrease can be seen in the maximum torque?

Difficulty level   ○ Easy   ● Normal   ○ Hard
Calculation amount   ○ Small   ● Normal   ○ Large

1) 19%
2) 20%
3) 29%
4) 10%

3.11. A three-phase wound-rotor induction motor has the stationary impedance of $0.05 + j\,0.2\ \Omega/phase$ referred to the stator side. What size of resistor must be added to each phase of rotor circuit so that the starting torque becomes maximum?

Difficulty level   ○ Easy   ● Normal   ○ Hard
Calculation amount   ○ Small   ● Normal   ○ Large

1) $0.25\ \Omega$
2) $0.1\ \Omega$
3) $0.15\ \Omega$
4) $0.5\ \Omega$

3.12. In a three-phase, four-pole, $50\ Hz$ induction motor, the speed of motor for the maximum torque is $750\ rpm$. Determine the ratio of starting torque to maximum torque of motor in percentage.

Difficulty level   ○ Easy   ● Normal   ○ Hard
Calculation amount   ○ Small   ● Normal   ○ Large

1) 90%
2) 70%
3) 80%
4) 120%

3.13. In an induction motor, the stationary impedance of rotor referred to the stator side is $1 + j1\,\Omega/phase$, but its stator impedance is negligible. This motor has the maximum torque three times as great compared to its full-load torque. Calculate the rated slip of motor.

Difficulty level   ○ Easy   ● Normal   ○ Hard
Calculation amount   ○ Small   ● Normal   ○ Large

1) 0.414
2) 0.268
3) 0.324
4) 0.171

3.14. In a three-phase, $50\ Hz$, $1440\ rpm$ induction motor, the full-load torque is twice of maximum torque. Determine the stable operating region of motor in terms of its speed.

Difficulty level   ○ Easy   ○ Normal   ● Hard
Calculation amount   ○ Small   ● Normal   ○ Large

1) $1080 - 1500\ rpm$
2) $1200 - 1500\ rpm$
3) $1180 - 1500\ rpm$
4) $1440 - 1500\ rpm$

3.15. A three-phase, eight-pole, $50\ Hz$, $700\ rpm$ induction motor has the stationary impedance of $0.01 + j\,0.05\ \Omega/phase$ referred to the stator side. What is the ratio of its full-load torque to maximum torque?

Difficulty level   ○ Easy   ○ Normal   ● Hard
Calculation amount   ○ Small   ● Normal   ○ Large

1) 1.8
2) 1.1
3) 0.6
4) 3.6

3.16. A three-phase, four-pole, 400 *V*, 50 *Hz* induction motor with a star-connected stator has the stationary impedance of $0.4 + j\,2\ \Omega/phase$ referred to the stator side. What size of resistor must be added to each phase of rotor circuit so that the starting torque becomes the 80% of maximum torque?
Difficulty level   ○ Easy   ○ Normal   ● Hard
Calculation amount   ○ Small   ● Normal   ○ Large
1) 0.8 Ω
2) 0.2 Ω
3) 0.4 Ω
4) 0.6 Ω

3.17. In a three-phase, four-pole, 380 *V*, 50 *Hz* induction motor, the value of parameters of equivalent circuit is as follows:

$$R_s = 0\ \Omega/phase, \quad R'_r = 0.5\ \Omega/phase$$

$$X_s = X'_r = 1\ \Omega/phase, \quad X_m = \infty\ \Omega/phase$$

Determine the speed in which the torque is maximum.
Difficulty level   ○ Easy   ○ Normal   ● Hard
Calculation amount   ○ Small   ● Normal   ○ Large
1) 1250 *rpm*
2) 1450 *rpm*
3) 1125 *rpm*
4) 1140 *rpm*

3.18. A three-phase, four-pole, 50 *Hz* wound-rotor induction motor is rotating under a certain load at 1440 *rpm*. After adding a series resistor to each phase of rotor circuit, its speed drops to about 1350 *rpm*. Calculate the ratio of ohmic power loss of rotor in this condition to the one of previous condition.
Difficulty level   ○ Easy   ○ Normal   ● Hard
Calculation amount   ○ Small   ● Normal   ○ Large
1) 2.5
2) 0.8
3) 1.25
4) 0.4

3.19. A four pole, 50 *Hz* induction motor has a negligible stator impedance. The rated speed of motor is 1475 *rpm*, and the stationary impedance of rotor is $0.072 + j\,0.8\ \Omega/phase$ referred to the stator side. Determine the unstable operating region of motor in terms of its speed.
Difficulty level   ○ Easy   ○ Normal   ● Hard
Calculation amount   ○ Small   ● Normal   ○ Large
1) 0 − 1365 *rpm*
2) 0 − 1450 *rpm*
3) 1450 − 1500 *rpm*
4) 1450 − 1475 *rpm*

3.20. In a three-phase, 50 *Hz*, 5 hp., 1440 *rpm* induction motor, the rotational power loss is 40 W. If the rotor current is 5 A, calculate the rotor resistance.
Difficulty level   ○ Easy   ○ Normal   ● Hard
Calculation amount   ○ Small   ● Normal   ○ Large
1) 1 Ω
2) 2 Ω
3) 3 Ω
4) 5 Ω

3.21. A three-phase, eight-pole, 50 *Hz* induction motor with the input power of 35 kW is rotating at 720 *rpm*. In this condition, the copper power loss of stator and rotational power loss are 1.5 kW and 160 W, respectively. Calculate the load torque while ignoring the core power loss.

Difficulty level   ○ Easy   ○ Normal   ● Hard
Calculation amount   ○ Small   ● Normal   ○ Large

1) $\frac{3000}{5\pi}$ *N.m*

2) $\frac{4000}{3\pi}$ *N.m*

3) $\frac{4000}{5\pi}$ *N.m*

4) $\frac{5000}{3\pi}$ *N.m*

3.22. A three-phase, six-pole, 50 *Hz* induction motor has the rotor impedance of $0.5 + j5$ *Ω/phase* referred to the stator side. Determine the speed in which the torque is maximum.

Difficulty level   ○ Easy   ○ Normal   ● Hard
Calculation amount   ○ Small   ● Normal   ○ Large

1) 1000 *rpm*

2) 1500 *rpm*

3) 900 *rpm*

4) 800 *rpm*

3.23. In a three-phase, two-pole, 50 *Hz*, induction motor, the starting torque is about 80% of rated torque. Calculate the motor speed at maximum torque.

Difficulty level   ○ Easy   ○ Normal   ● Hard
Calculation amount   ○ Small   ● Normal   ○ Large

1) 900 *rpm*

2) 1000 *rpm*

3) 1200 *rpm*

4) 1500 *rpm*

3.24. In a squirrel-cage induction motor, if the copper bars are replaced by the similar aluminum bars, calculate the ratio of speeds corresponding to maximum torque ($\frac{n_{T \, max \, .Al}}{n_{T \, max \, .Cu}}$). Herein, assume that the special electrical conductivity of copper is two times as great compared to the one of aluminum.

Difficulty level   ○ Easy   ○ Normal   ● Hard
Calculation amount   ○ Small   ● Normal   ○ Large

1) 1

2) Less than one

3) 2

4) More than 2

3.25. A three-phase, four-pole, 50 *Hz* induction motor is rotating at the speed of 1450 *rpm*. If the rotor resistance becomes three times as great, what will be the new speed of machine?

Difficulty level   ○ Easy   ○ Normal   ● Hard
Calculation amount   ○ Small   ● Normal   ○ Large

1) 1450 *rpm*

2) 1400 *rpm*

3) 1350 *rpm*

4) 1500 *rpm*

3.26. A three-phase, six-pole, 50 *Hz* induction motor has the rated speed of 950 *rpm*. What is its speed at half-load condition if the motor is still being operated in the stable region?

Difficulty level   ○ Easy   ○ Normal   ● Hard
Calculation amount   ○ Small   ● Normal   ○ Large

1) 975 *rpm*

2) 1000 *rpm*

3) 980 *rpm*

4) 950 *rpm*

3.27. A three-phase, six-pole, 50 *Hz* induction motor develops the net torque of 120 N.m. when the frequency of electromotive force (emf) of rotor is about 2 *Hz*. Calculate the efficiency of machine if the rotational power loss and the total power loss of stator are about 2 kW and 0.5 kW, respectively.

Difficulty level   ○ Easy   ○ Normal   ● Hard

Calculation amount   ○ Small   ○ Normal   ● Large

1) 84.2%

2) 82.8%

3) 79.6%

4) 76.7%

3.28. A three-phase, four-pole, 50 *Hz* induction motor is supplied by a 500 *V* power source and rotates at 1440 *rpm*. The output power of rotor is about 20 hp. Calculate the copper power loss of rotor.

Difficulty level   ○ Easy   ○ Normal   ● Hard

Calculation amount   ○ Small   ○ Normal   ● Large

1) 425.45 W

2) 521.32 W

3) 575.65 W

4) 621.66 W

3.29. In a three-phase, six-pole, 50 *Hz*, 12 hp., 950 *rpm* induction motor, the copper power loss of stator and the core power loss are about 1 kW and 0.5 kW, respectively. Calculate the efficiency of machine if its rotational power loss is negligible.

Difficulty level   ○ Easy   ○ Normal   ● Hard

Calculation amount   ○ Small   ○ Normal   ● Large

1) 85.0%

2) 82.0%

3) 82.6%

4) 85.8%

3.30. A three-phase, four-pole, 50 *Hz* induction motor has the efficiency of 85% at the power of 25.5 kW. If the copper power loss of stator, copper power loss of rotor, core power loss of machine, and rotational power loss are equal, calculate the speed of machine.

Difficulty level   ○ Easy   ○ Normal   ● Hard

Calculation amount   ○ Small   ○ Normal   ● Large

1) 1425 *rpm*

2) 1438 *rpm*

3) 1440 *rpm*

4) 1500 *rpm*

3.31. In a three-phase, eight-pole, 50 *Hz*, 700 *rpm* induction motor, the input power is 5 kW. Calculate the efficiency of motor if the total power loss of stator and rotational power loss are 100 W and 150 W, respectively.

Difficulty level   ○ Easy   ○ Normal   ● Hard

Calculation amount   ○ Small   ○ Normal   ● Large

1) 78.4%

2) 85.4%

3) 92.4%

4) 88.4%

3.32. A three-phases 50 *Hz* induction motor can be operated with two and four poles. While the machine is operated with two poles and the slip of 0.08, it is changed to a four-pole machine. Calculate the ratio of developed electromagnetic power to input power immediately after this switching. Herein, ignore the stator power loss.

Difficulty level   ○ Easy   ○ Normal   ● Hard
Calculation amount   ○ Small   ○ Normal   ● Large

1) 0.16
2) 1.84
3) 1.08
4) −1.84

3.33. A three-phase, six-pole, 50 *Hz*, 50 kW, 970 *rpm* induction motor has the core power loss, stator copper power loss, and rotational power loss of 1 kW, 1 kW, and 1.5 kW, respectively. Calculate the efficiency of machine.

Difficulty level   ○ Easy   ○ Normal   ● Hard
Calculation amount   ○ Small   ○ Normal   ● Large

1) 91%
2) 81%
3) 85%
4) 71%

3.34. A three-phase, four-pole, 9.5 kW, 1455 *rpm*, 50 *Hz* induction motor is available. The rotational power loss of machine at the rated speed is 200 W. If the rotor current is 5 A, calculate the rotor resistance.

Difficulty level   ○ Easy   ○ Normal   ● Hard
Calculation amount   ○ Small   ○ Normal   ● Large

1) 11.5 *Ω/phase*
2) 12 *Ω/phase*
3) 3.8 *Ω/phase*
4) 4 *Ω/phase*

3.35. A four-pole, 60 *Hz*, wound-rotor induction motor has been loaded with a constant-torque load and rotates at the rated speed of 1710 *rpm*. Calculate the ratio of developed electromagnetic power to air gap power if the air gap magnetic flux and rotational power loss remain constant but the rotor resistance becomes five times as great.

Difficulty level   ○ Easy   ○ Normal   ● Hard
Calculation amount   ○ Small   ○ Normal   ● Large

1) 0.62
2) 0.75
3) 0.82
4) 0.90

3.36. A three-phase, four-pole, 50 *Hz* induction motor, at its rated voltage and frequency, has the starting torque equal to one and a half times of the rated torque as well as the maximum torque equal to three times of the rated torque. Calculate its rated speed.

Difficulty level   ○ Easy   ○ Normal   ● Hard
Calculation amount   ○ Small   ○ Normal   ● Large

1) 1431 *rpm*
2) 1350 *rpm*
3) 1200 *rpm*
4) 1098 *rpm*

3.37. In a 50 *Hz*, 1425 *rpm*, induction motor, the speed is controlled by the method of $\frac{V}{f}$ = Constant for the loads with constant torque. If both voltage and frequency become one and a half as great, what will be the new speed of motor? Herein, ignore the stator impedance.

Difficulty level   ○ Easy   ○ Normal   ● Hard

Calculation amount   ○ Small   ○ Normal   ● Large

1) 2137.5 *rpm*

2) 2175.0 *rpm*

3) 2112.5 *rpm*

4) 2237.5 *rpm*

Solutions of Problems: Induction Machines

4

Abstract

In this chapter, the problems of the third chapter are fully solved, in detail, step-by-step, and with different methods.

4.1. Since the load is not changed, the line current will be the same as the previous one. In other words, changing the connection type of stator winding will not affect the current drawn from the grid. Choice (1) is the answer.

However, it should be noted that in delta connection, the phase current will be $\frac{1}{\sqrt{3}}$ times of line current.

4.2. Based on the information given in the problem, we have:

$$p = 6 \tag{1}$$

$$f_s = 50 \, Hz \tag{2}$$

$$f_r = 100 \, cycle/min \Rightarrow f_r = \frac{100}{60} \, cycle/sec = \frac{5}{3} \, Hz \tag{3}$$

As we know, the slip can be calculated as follows:

$$s = \frac{f_r}{f_s} \tag{4}$$

$$\Rightarrow s = \frac{\frac{5}{3}}{50} \Rightarrow s \approx 0.033 \tag{5}$$

The synchronous speed can be calculated based on the number of poles of rotor and frequency of stator as follows:

$$n_s = \frac{120 f_s}{p} \tag{6}$$

$$\Rightarrow n_s = \frac{120 \times 50}{6} \Rightarrow n_s = 1000 \, rpm \tag{7}$$

The speed of rotor can be determined as follows:

$$n = n_s(1 - s) \tag{8}$$

$$\Rightarrow n = 1000(1 - 0.033) \Rightarrow n \approx 966 \, rpm$$

Choice (4) is the answer.

© The Author(s), under exclusive license to Springer Nature Switzerland AG 2023
M. Rahmani-Andebili, *AC Electric Machines*, https://doi.org/10.1007/978-3-031-15139-2_4

4.3. Based on the information given in the problem, we have:

$$V_1 = 380 \ V \tag{1}$$

$$T_{st,1} = 1.5 T_{FL} \tag{2}$$

$$T_{st,2} = T_{FL} \tag{3}$$

As we know, at constant frequency, the starting torque is proportional to squire of voltage:

$$T_{st} \propto V^2 \tag{4}$$

Therefore:

$$\frac{T_{st,1}}{T_{st,2}} = \left(\frac{V_1}{V_2}\right)^2 \tag{5}$$

$$\Rightarrow 1.5 = \left(\frac{380}{V_2}\right)^2 \Rightarrow V_2 \approx 310 \ V$$

Choice (2) is the answer.

4.4. Based on the information given in the problem, we have:

$$V_1 = 220 \ V \tag{1}$$

$$T_{max,1} = 3 T_{FL} \tag{2}$$

$$V_2 = 110 \ V \tag{3}$$

As we know, at constant frequency, the maximum torque is proportional to squire of voltage:

$$T_{max} \propto V^2 \tag{4}$$

Therefore:

$$\frac{T_{max,1}}{T_{max,2}} = \left(\frac{220}{110}\right)^2 \tag{5}$$

$$\Rightarrow T_{max,2} = \frac{1}{4} T_{max,1} \tag{6}$$

Solving (2) and (6):

$$\Rightarrow T_{max,2} = \frac{3}{4} T_{FL} \Rightarrow \frac{T_{max,2}}{T_{FL}} = 0.75$$

Choice (2) is the answer.

4.5. Based on the information given in the problem, we have:

$$p = 4 \tag{1}$$

$$f_s = 50 \ Hz \tag{2}$$

$$s = 0.05 \tag{3}$$

The synchronous speed can be calculated based on the number of poles of rotor and frequency of stator as follows:

$$n_s = \frac{120 f_s}{p} \tag{4}$$

$$\Rightarrow n_s = \frac{120 \times 50}{4} \Rightarrow n_s = 1500 \ rpm \tag{5}$$

The speed of rotor can be determined as follows:

$$n = n_s(1 - s) \tag{6}$$

$$\Rightarrow n = 1500(1 - 0.05) \Rightarrow n = 1425 \ rpm \tag{7}$$

As we know, speed of revolving field of rotor and stator is equal. In other words:

$$n_r = n_s \tag{8}$$

The speed of revolving field of rotor or stator with respect to rotor can be calculated as follows:

$$n_s - n = n_r - n = 1500 - 1425 = 75 \ rpm$$

Choice (4) is the answer.

4.6. Based on the information given in the problem, we have:

$$p = 6 \tag{1}$$

$$f_s = 60 \ Hz \tag{2}$$

$$n = 1140 \ rpm \tag{3}$$

The synchronous speed can be calculated based on the number of poles of rotor and frequency of stator as follows:

$$n_s = \frac{120 f_s}{p} \tag{4}$$

$$\Rightarrow n_s = \frac{120 \times 60}{4} \Rightarrow n_s = 1200 \ rpm \tag{5}$$

As we know, the slip can be calculated as follows:

$$s = \frac{n_s - n}{n_s} \tag{6}$$

$$\Rightarrow s = \frac{1200 - 1140}{1200} \Rightarrow s = 0.05 \tag{7}$$

The frequency of rotor is calculated as follows:

$$f_r = sf_s \Rightarrow f_r = 0.05 \times 60 \Rightarrow f_r = 3\,Hz$$

Choice (4) is the answer.

4.7. Based on the information given in the problem, we have:

$$P_{in} = 4000\,W \tag{1}$$

$$P_{cu,s} + P_c = 100\,W \tag{2}$$

$$P_{rot} = 200\,W \tag{3}$$

$$s = 0.04 \tag{4}$$

Figure 4.1 shows the power flow diagram of an induction motor. The air gap power can be calculated as follows:

$$P_{ag} = P_{in} - (P_{cu,s} + P_c) = 4000 - 100 = 3900\,W \tag{5}$$

As we know, the developed electromagnetic power can be calculated as follows:

$$P_e = (1-s)P_{ag} = (1-0.04) \times 3900 = 3744\,W \tag{6}$$

Based on the power flow diagram, the out power is calculated as follows:

$$P_{out} = P_e - P_{rot} = 3744 - 200 = 3544\,W \tag{7}$$

The efficiency of motor can be calculated as follows:

$$\eta = \frac{P_{out}}{P_{in}} \times 100 \tag{8}$$

$$\Rightarrow \eta = \frac{3544}{4000} \times 100 \Rightarrow \eta = 88.6\%$$

Choice (2) is the answer.

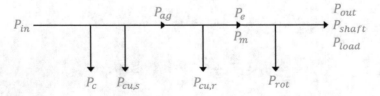

Fig. 4.1 The circuit of solution of Problem 4.7

4.8. Based on the information given in the problem, we have:

$$p = 8 \tag{1}$$

$$f_s = 60\,Hz \tag{2}$$

$$P_{ag} = 90\,kW \tag{3}$$

$$\omega_s - \omega = 10 \; rad/sec \tag{4}$$

$$P_{rot} \approx 0 \; W \tag{5}$$

As we know, the synchronous speed can be calculated based on the number of poles of rotor and frequency of stator as follows:

$$n_s = \frac{120 f_s}{p} \tag{6}$$

$$\Rightarrow n_s = \frac{120 \times 60}{8} \Rightarrow n_s = 900 \; rpm \tag{7}$$

$$\Rightarrow \omega_s = 900 \times \frac{2\pi}{60} = 30\pi \; rad/sec \tag{8}$$

As we know, the relation below exists:

$$\omega = \omega_s(1 - s) \Rightarrow \omega - \omega_s = -\omega_s s \tag{9}$$

Solving (4), (8), and (9):

$$-10 = -30\pi s \Rightarrow s = \frac{1}{3\pi} \tag{10}$$

As we know, the developed electromagnetic power can be calculated as follows:

$$P_e = (1 - s)P_{ag} = \left(1 - \frac{1}{3\pi}\right) \times 90 \Rightarrow P_e \approx 80.45 \; kW$$

Finally, the output power can be calculated as follows:

$$P_{out} = P_e - P_{rot} = 80.45 - 0 \Rightarrow P_{out} \approx 80.45 \; kW$$

Choice (2) is the answer.

4.9. Based on the information given in the problem, we have:

$$Z'_r = 0.05 + j0.1 \; \Omega/phase \tag{1}$$

$$T_{st} = T_{max} \tag{2}$$

As we know, the slip corresponding to maximum torque is as follows if only rotor impedance is considerable:

$$s_{T_{max}} = \frac{R'_r}{X'_r} = \frac{R_r}{X_r} \tag{3}$$

If an external resistor is added to the rotor circuit, the slip corresponding to maximum torque is updated as follows:

$$s_{T_{max}} = \frac{R_r + R_{ext}}{X_r} \tag{4}$$

Since the maximum torque needs to occur during the starting time, $s_{T_{max}} = 1$. Therefore:

$$1 = \frac{R_r + R_{ext}}{X_r} \tag{5}$$

Solving (1) and (5):

$$1 = \frac{0.05 + R_{ext}}{0.1} \Rightarrow R_{ext} = 0.05 \ \Omega/phase$$

Choice (4) is the answer.

4.10. Based on the information given in the problem, we have:

$$V_2 = 0.9 V_1 \tag{1}$$

As we know, at constant frequency, the maximum torque is proportional to squire of voltage:

$$T_{max} \propto V^2 \tag{2}$$

Therefore:

$$\frac{T_{max,2}}{T_{max,1}} = \left(\frac{V_2}{V_1}\right)^2 \tag{3}$$

$$\Rightarrow \frac{T_{max,2}}{T_{max,1}} = (0.9)^2 \Rightarrow T_{max,2} = 0.81 T_{max,1} \tag{4}$$

$$\Rightarrow \frac{T_{max,2} - T_{max,1}}{T_{max,1}} = -0.19 \Rightarrow 19\% \ \text{Reduction}$$

Choice (1) is the answer.

4.11. Based on the information given in the problem, we have:

$$Z'_r = 0.05 + j0.2 \ \Omega/phase \tag{1}$$

$$T_{st} = T_{max} \tag{2}$$

As we know, the slip corresponding to maximum torque is as follows if only rotor impedance is considerable:

$$s_{T_{max}} = \frac{R'_r}{X'_r} = \frac{R_r}{X_r} \tag{3}$$

If an external resistor is added to the rotor circuit, the slip corresponding to maximum torque is updated as follows:

$$s_{T_{max}} = \frac{R_r + R_{ext}}{X_r} \tag{4}$$

Since the maximum torque needs to occur during the starting time, $s_{T_{max}} = 1$. Therefore:

$$1 = \frac{R_r + R_{ext}}{X_r} \tag{5}$$

Solving (1) and (5):

$$1 = \frac{0.05 + R_{ext}}{0.2} \Rightarrow R_{ext} = 0.15 \ \Omega/phase$$

Choice (3) is the answer.

4.12. Based on the information given in the problem, we have:

$$p = 4 \tag{1}$$

$$f_s = 50 \ Hz \tag{2}$$

$$n_{T_{max}} = 750 \ rpm \tag{3}$$

The synchronous speed can be calculated based on the number of poles of rotor and frequency of stator as follows:

$$n_s = \frac{120 f_s}{p} \tag{4}$$

$$\Rightarrow n_s = \frac{120 \times 50}{4} \Rightarrow n_s = 1500 \ rpm \tag{5}$$

The slip at maximum torque can be calculated as follows:

$$s_{T_{max}} = \frac{n_s - n_{T_{max}}}{n_s} \tag{6}$$

$$\Rightarrow s_{T_{max}} = \frac{1500 - 750}{1500} \Rightarrow s_{T_{max}} = 0.5 \tag{7}$$

As we know, the relation between starting torque and maximum torque is as follows:

$$\frac{T_{st}}{T_{max}} = \frac{2 s_{T_{max}}}{\left(s_{T_{max}}\right)^2 + 1} \tag{8}$$

$$\Rightarrow \frac{T_{st}}{T_{max}} = \frac{2 \times 0.5}{(0.5)^2 + 1} \Rightarrow \frac{T_{st}}{T_{max}} = 0.8 = 80\%$$

Choice (3) is the answer.

4.13. Based on the information given in the problem, we have:

$$Z'_r = 1 + j \ \Omega/phase \tag{1}$$

$$Z_s \approx 0 \ \Omega/phase \tag{2}$$

$$T_{max} = 3 T_{FL} \tag{3}$$

As we know, the slip corresponding to maximum torque is as follows if only rotor impedance is considerable:

$$s_{T_{max}} = \frac{R'_r}{X'_r} = \frac{R_r}{X_r} \tag{4}$$

$$\Rightarrow s_{T_{max}} = \frac{1}{1} = 1 \tag{5}$$

The relation between full-load torque and maximum torque is as follows:

$$\frac{T_{FL}}{T_{max}} = \frac{2s_{T_{max}}s_{T_{FL}}}{\left(s_{T_{max}}\right)^2 + \left(s_{T_{FL}}\right)^2} \tag{6}$$

$$\Rightarrow \frac{1}{3} = \frac{2 \times 1 \times s_{T_{FL}}}{(1)^2 + \left(s_{T_{FL}}\right)^2} \Rightarrow s_{T_{FL}} = \begin{cases} 5.82 \Rightarrow \text{Uncceptable} \\ 0.171 \Rightarrow \text{Acceptable} \end{cases} \tag{7}$$

Since the machine is operated as a motor $(0 < s < 1)$, only $s_{T_{FL}} = 0.171$ is acceptable.

Choice (4) is the answer.

4.14. Based on the information given in the problem, we have:

$$f_s = 50 \ Hz \tag{1}$$

$$n_{FL} = 1440 \ rpm \tag{2}$$

$$T_{FL} = 2T_{max} \tag{3}$$

Based on the speed of motor given in the problem, we can determine the synchronous speed of machine as follows:

$$n_s = \frac{120f_s}{p} \tag{4}$$

$$\Rightarrow n_s = \frac{120 \times 50}{p} = \frac{6000}{p} \Rightarrow \begin{cases} p = 1 \Rightarrow n_s = 6000 \\ p = 2 \Rightarrow n_s = 3000 \\ p = 4 \Rightarrow n_s = 1500 \\ p = 6 \Rightarrow n_s = 1000 \\ p = 8 \Rightarrow n_s = 7500 \end{cases} \tag{5}$$

Since speed of an induction machine is close to its synchronous speed during stable operation (rated condition), it is clear that $n_s = 1500$ for this machine because $n = 1440 \ rpm$.

The slip at full-load condition can be calculated as follows:

$$s_{FL} = \frac{n_s - n_{FL}}{n_s} \tag{6}$$

$$\Rightarrow s_{FL} = \frac{1500 - 1440}{1500} \Rightarrow s_{FL} = 0.04 \tag{7}$$

As we know, the relation between full-load torque and maximum torque is as follows:

$$\frac{T_{FL}}{T_{max}} = \frac{2s_{T_{max}}s_{FL}}{\left(s_{T_{max}}\right)^2 + \left(s_{FL}\right)^2} \tag{8}$$

$$\Rightarrow 2 = \frac{2s_{T_{max}} \times 0.04}{\left(s_{T_{max}}\right)^2 + (0.04)^2} \Rightarrow s_{T_{max}} = 0.28 \tag{9}$$

The speed of motor corresponding to the maximum torque can be calculated as follows:

$$n_{T_{max}} = n_s(1 - s_{T_{max}}) \tag{10}$$

$$\Rightarrow n_{T_{max}} = 1500(1 - 0.28) \Rightarrow n_{T_{max}} = 1080 \ rpm \tag{11}$$

The speed corresponding to maximum torque is the minimum speed in which the induction motor has its own stable operation. Therefore, the speeds above the minimum speed, but below the synchronous speed, will result in the stable operation of motor. In other words, the stable operating region of motor is as follows:

$$1080 < n < 1500 \ rpm$$

Choice (1) is the answer.

4.15. Based on the information given in the problem, we have:

$$p = 8 \tag{1}$$

$$f_s = 50 \ Hz \tag{2}$$

$$n = 700 \ rpm \tag{3}$$

$$Z_r' = 0.01 + j0.05 \ \Omega/phase \tag{4}$$

The synchronous speed can be calculated based on the number of poles of rotor and frequency of stator as follows:

$$n_s = \frac{120f_s}{p} \tag{5}$$

$$\Rightarrow n_s = \frac{120 \times 50}{8} \Rightarrow n_s = 750 \ rpm \tag{6}$$

The slip at full-load condition can be calculated as follows:

$$s_{T_{FL}} = \frac{n_s - n_{T_{FL}}}{n_s} \tag{7}$$

$$\Rightarrow s_{T_{FL}} = \frac{750 - 700}{750} \Rightarrow s_{T_{FL}} = \frac{1}{15} \tag{8}$$

As we know, the slip corresponding to maximum torque is as follows if only rotor impedance is considerable:

$$s_{T_{max}} = \frac{R_r'}{X_r'} = \frac{R_r}{X_r} \tag{9}$$

$$s_{T_{max}} = \frac{0.01}{0.05} = 0.2 \tag{10}$$

The relation between full-load torque and maximum torque is as follows:

$$\frac{T_{FL}}{T_{max}} = \frac{2s_{T_{max}}s_{T_{FL}}}{\left(s_{T_{max}}\right)^2 + \left(s_{T_{FL}}\right)^2} \tag{11}$$

$$\Rightarrow \frac{T_{FL}}{T_{max}} = \frac{2 \times 0.2 \times \frac{1}{15}}{(0.2)^2 + \left(\frac{1}{15}\right)^2} \Rightarrow \frac{T_{FL}}{T_{max}} = 0.6$$

Choice (3) is the answer.

4.16. Based on the information given in the problem, we have:

$$p = 4 \tag{1}$$

$$f_s = 50 \, Hz \tag{2}$$

$$Z'_r = 0.4 + j2 \, \Omega/phase \tag{3}$$

$$T_{st} = 0.8 T_{max} \tag{4}$$

As we know, the relation between starting torque and maximum torque is as follows:

$$\frac{T_{st}}{T_{max}} = \frac{2 s_{T_{max}}}{\left(s_{T_{max}}\right)^2 + 1} \tag{5}$$

$$\Rightarrow 0.8 = \frac{2 s_{T_{max}}}{\left(s_{T_{max}}\right)^2 + 1} \Rightarrow s_{T_{max}} = \begin{cases} 2 \Rightarrow \text{Uncceptable} \\ 0.5 \Rightarrow \text{Acceptable} \end{cases} \tag{6}$$

Since the machine is operated as a motor ($0 < s < 1$), only $s_{T_{max}} = 0.5$ is acceptable.

As we know, the slip corresponding to maximum torque is as follows if only rotor impedance is considerable:

$$s_{T_{max}} = \frac{R'_r}{X'_r} = \frac{R_r}{X_r} \tag{7}$$

If an external resistor is added to the rotor circuit, the slip corresponding to maximum torque is updated as follows:

$$s_{T_{max}} = \frac{R_r + R_{ext}}{X_r} \tag{8}$$

Therefore:

$$0.5 = \frac{0.4 + R_{ext}}{2} \Rightarrow R_{ext} = 0.6 \, \Omega/phase$$

Choice (4) is the answer.

4.17. Based on the information given in the problem, we have:

$$p = 4 \tag{1}$$

$$f_s = 50 \, Hz \tag{2}$$

$$R_s = 0 \, \Omega/phase, R'_r = 0.5 \, \Omega/phase \tag{3}$$

$$X_s = X'_r = 1 \, \Omega/phase, X_m = \infty \ \Omega/phase \tag{4}$$

As we know, the relation between Thevenin impedance and the other parameters of induction machine is as follows:

$$R_{th} = \left(\frac{X_m}{X_s + X_m}\right)^2 R_s \qquad (5)$$

$$X_{th} \approx X_s \qquad (6)$$

Therefore:

$$R_{th} = 0 \ \Omega/phase \qquad (7)$$

$$X_{th} \approx 1 \ \Omega/phase \qquad (8)$$

The slip corresponding to maximum torque can be calculated as follows:

$$s_{T_{max}} = \frac{R'_r}{\sqrt{(R_{th})^2 + (X_{th} + X'_r)^2}} \qquad (9)$$

$$\Rightarrow s_{T_{max}} = \frac{0.5}{\sqrt{(0)^2 + (1+1)^2}} = 0.25 \qquad (10)$$

The synchronous speed can be calculated based on the number of poles of rotor and frequency of stator as follows:

$$n_s = \frac{120 f_s}{p} \qquad (11)$$

$$\Rightarrow n_s = \frac{120 \times 50}{4} \Rightarrow n_s = 1500 \ rpm \qquad (12)$$

The speed of motor corresponding to the maximum torque can be calculated as follows:

$$n_{T_{max}} = n_s (1 - s_{T_{max}}) \qquad (13)$$

$$\Rightarrow n_{T_{max}} = 1500(1 - 0.25) \Rightarrow n_{T_{max}} = 1125 \ rpm$$

Choice (3) is the answer.

4.18. Based on the information given in the problem, we have:

$$p = 4 \qquad (1)$$

$$f_s = 50 \ Hz \qquad (2)$$

$$n_1 = 1440 \ rpm \qquad (3)$$

$$n_2 = 1350 \ rpm \qquad (4)$$

The synchronous speed can be calculated based on the number of poles of rotor and frequency of stator as follows:

$$n_s = \frac{120 f_s}{p} \qquad (5)$$

$$\Rightarrow n_s = \frac{120 \times 50}{4} \Rightarrow n_s = 1500 \ rpm \tag{6}$$

The slip in the first condition can be calculated as follows:

$$s_1 = \frac{n_s - n_1}{n_s} \tag{7}$$

$$\Rightarrow s_1 = \frac{1500 - 1440}{1500} \Rightarrow s_1 = 0.04 \tag{8}$$

Likewise for the second condition, we have:

$$\Rightarrow s_2 = \frac{1500 - 1350}{1500} \Rightarrow s_2 = 0.1 \tag{9}$$

As we know, the relation between copper power loss of rotor and air gap power is as follows:

$$P_{cu,r} = sP_{ag} \tag{10}$$

Therefore:

$$\frac{P_{cu,r,2}}{P_{cu,r,1}} = \frac{s_2 P_{ag,2}}{s_1 P_{ag,1}} \tag{11}$$

Since the load torque and synchronous speed are constant, the air gap power will not change based on $P_{ag} = T_e \omega_s$. In other words:

$$P_{ag,2} = P_{ag,1} \tag{12}$$

Solving (11) and (12) and considering the value of parameters:

$$\frac{P_{cu,r,2}}{P_{cu,r,1}} = \frac{s_2}{s_1} = \frac{0.1}{0.04} \Rightarrow \frac{P_{cu,r,2}}{P_{cu,r,1}} = 2.5$$

Choice (1) is the answer.

4.19. Based on the information given in the problem, we have:

$$p = 4 \tag{1}$$

$$f_s = 50 \ Hz \tag{2}$$

$$Z_s \approx 0 \ \Omega/phase \tag{3}$$

$$Z_r' = 0.072 + j0.8 \ \Omega/phase \tag{4}$$

As we know, the slip corresponding to maximum torque is as follows if only rotor impedance is considerable:

$$s_{T_{max}} = \frac{R_r'}{X_r'} = \frac{R_r}{X_r} \tag{5}$$

$$\Rightarrow s_{T_{max}} = \frac{0.072}{0.8} = 0.09 \tag{6}$$

The synchronous speed can be calculated based on the number of poles of rotor and frequency of stator as follows:

$$n_s = \frac{120 f_s}{p} \tag{7}$$

$$\Rightarrow n_s = \frac{120 \times 50}{4} \Rightarrow n_s = 1500 \ rpm \tag{8}$$

The speed of motor corresponding to the maximum torque can be calculated as follows:

$$n_{T_{max}} = n_s (1 - s_{T_{max}}) \tag{9}$$

$$\Rightarrow n_{T_{max}} = 1500(1 - 0.09) \Rightarrow n_{T_{max}} = 1365 \ rpm \tag{10}$$

The speed corresponding to maximum torque is the minimum speed in which the induction motor has its own stable operation. Therefore, the speeds below this value will result in the unstable operation of motor. In other words, $n_{min} = 1365 \ rpm$. Thus, the unstable operating region of motor is as follows:

$$0 < n < 1365 \ rpm$$

Choice (1) is the answer.

4.20. Based on the information given in the problem, we have:

$$f_s = 50 \ Hz \tag{1}$$

$$P_{out} = 5 \ hp \tag{2}$$

$$n = 1440 \ rpm \tag{3}$$

$$P_{rot} = 40 \ W \tag{4}$$

$$I_r = 5 \ A \tag{5}$$

The new slip can be calculated as follows:

$$s = \frac{n_s - n}{n_s} \tag{6}$$

$$\Rightarrow s = \frac{1500 - 1440}{1500} \Rightarrow s = 0.04 \tag{7}$$

Figure 4.2 shows the power flow diagram of an induction motor. The electromagnetic power can be calculated as follows:

$$P_e = P_{out} + P_{rot} \tag{8}$$

$$\Rightarrow P_e = 5 \times 746 + 40 = 3770 \ W \tag{9}$$

As we know, the copper power loss of rotor can be calculated as follows:

$$P_{cu,r} = \frac{s}{1-s}P_e = \frac{0.04}{1-0.04} \times 3770 = 157\ W \tag{10}$$

On the other hand, the copper power loss of rotor of a three-phase induction motor can be directly calculated as follows:

$$P_{cu,r} = 3R_r(I_r)^2 \tag{11}$$

$$\Rightarrow P_{cu,r} = 3R_r(5)^2 = 75R_r \tag{12}$$

Solving (10) and (12):

$$R_r = \frac{157}{75} \Rightarrow R_r \approx 2\ \Omega/phase$$

Choice (2) is the answer.

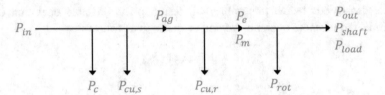

Fig. 4.2 The circuit of solution of Problem 4.20

4.21. Based on the information given in the problem, we have:

$$p = 8 \tag{1}$$

$$f_s = 50\ Hz \tag{2}$$

$$P_{in} = 35\ kW \tag{3}$$

$$n = 720\ rpm \tag{4}$$

$$P_{cu,s} = 1.5\ kW \tag{5}$$

$$P_{rot} = 0.16\ kW \tag{6}$$

$$P_c = 0\ kW \tag{7}$$

The synchronous speed can be calculated based on the number of poles of rotor and frequency of stator as follows:

$$n_s = \frac{120f_s}{p} \tag{8}$$

$$\Rightarrow n_s = \frac{120 \times 50}{8} \Rightarrow n_s = 750\ rpm \tag{9}$$

The slip can be calculated as follows:

$$s = \frac{n_s - n}{n_s} \tag{10}$$

$$\Rightarrow s = \frac{750 - 720}{750} \Rightarrow s = 0.04 \tag{11}$$

Figure 4.3 shows the power flow diagram of an induction motor. The air gap power can be calculated as follows:

$$P_{ag} = P_{in} - (P_{cu,s} + P_c) = 35 - (1.5 + 0) = 33.5 \, kW \tag{12}$$

As we know, the developed electromagnetic power can be calculated as follows:

$$P_e = (1 - s)P_{ag} = (1 - 0.04) \times 33.5 = 32.16 \, kW \tag{13}$$

Based on the power flow diagram, the output power can be calculated as follows:

$$P_{out} = P_e - P_{rot} = 32.16 - 0.16 = 32 \, kW \tag{14}$$

As we know, the output torque can be calculated as follow

$$T_{out} = \frac{P_{out}}{\omega} = \frac{P_{out}}{\frac{2\pi}{60} \times n} \tag{15}$$

$$\Rightarrow T_{out} = \frac{32000}{\frac{2\pi}{60} \times 720} \Rightarrow T_{out} = \frac{4000}{3\pi} \, N.m$$

Choice (2) is the answer.

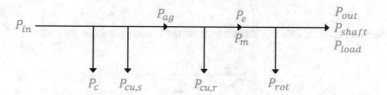

Fig. 4.3 The circuit of solution of Problem 4.21

4.22. Based on the information given in the problem, we have:

$$p = 6 \tag{1}$$

$$f_s = 50 \, Hz \tag{2}$$

$$Z'_r = 0.5 + j5 \, \Omega/phase \tag{3}$$

The slip corresponding to maximum torque is as follows:

$$s_{T_{max}} = \frac{R'_r}{\sqrt{(R_{th})^2 + (X_{th} + X'_r)^2}} \tag{4}$$

The impedances of stator and excitation branch are not given in the problem. It means that the impedance of stator is almost zero and the impedance of excitation branch is too large. Therefore, the Thevenin impedance seen by the rotor is ignorable ($R_{th} = X_{th} = 0$). Therefore, we have:

$$s_{T_{max}} = \frac{R'_r}{X'_r} = \frac{R_r}{X_r} \tag{5}$$

$$\Rightarrow s_{T_{max}} = \frac{0.5}{5} = 0.1 \tag{6}$$

The synchronous speed can be calculated based on the number of poles of rotor and frequency of stator as follows:

$$n_s = \frac{120 f_s}{p} \tag{7}$$

$$\Rightarrow n_s = \frac{120 \times 50}{6} \Rightarrow n_s = 1000 \; rpm \tag{8}$$

Now, the speed of motor corresponding to the maximum torque can be calculated as follows:

$$n_{T_{max}} = n_s(1 - s_{T_{max}}) \tag{9}$$

$$\Rightarrow n_{T_{max}} = 1000(1 - 0.1) \Rightarrow n_{T_{max}} = 900 \; rpm$$

Choice (3) is the answer.

4.23. Based on the information given in the problem, we have:

$$p = 2 \tag{1}$$

$$f_s = 50 \; Hz \tag{2}$$

$$T_{st} = 0.8 T_{FL} \tag{3}$$

As we know, the relation between starting torque and maximum torque is as follows:

$$\frac{T_{st}}{T_{max}} = \frac{2 s_{T_{max}}}{\left(s_{T_{max}}\right)^2 + 1} \tag{4}$$

$$\Rightarrow 0.8 = \frac{2 s_{T_{max}}}{\left(s_{T_{max}}\right)^2 + 1} \Rightarrow s_{T_{max}} = \begin{cases} 2 \Rightarrow \text{Uncceptable} \\ 0.5 \Rightarrow \text{Acceptable} \end{cases} \tag{5}$$

Since the machine is operated as a motor $(0 < s < 1)$, only $s_{T_{max}} = 0.5$ is acceptable.

The synchronous speed can be calculated based on the number of poles of rotor and frequency of stator as follows:

$$n_s = \frac{120 f_s}{p} \tag{6}$$

$$\Rightarrow n_s = \frac{120 \times 50}{2} \Rightarrow n_s = 3000 \; rpm \tag{7}$$

The speed of motor corresponding to the maximum torque can be calculated as follows:

$$n_{T_{max}} = n_s(1 - s_{T_{max}}) \tag{8}$$

$$\Rightarrow n_{T_{max}} = 3000(1 - 0.5) \Rightarrow n_{T_{max}} = 1500 \; rpm$$

Choice (4) is the answer.

4.24. Based on the information given in the problem, we have:

$$\sigma_{Cu} \triangleq 2\sigma_{Al} \tag{1}$$

From (1), it is concluded that:

$$R'_{Al} = 2R'_{Cu} \tag{2}$$

$$\xrightarrow{\times \frac{1}{X'_r}} \frac{R'_{r,Al}}{X'_r} = 2\frac{R'_{r,Cu}}{X'_r} \tag{3}$$

As we know, the slip corresponding to maximum torque is as follows if only rotor impedance is considerable:

$$s_{T_{max}} = \frac{R'_r}{X'_r} = \frac{R_r}{X_r} \tag{4}$$

Solving (3) and (4):

$$s_{T_{max},Al} = 2s_{T_{max},Cu} \tag{5}$$

The ratio of speeds of motor in the two conditions can be calculated as follows:

$$\frac{n_{T_{max},Al}}{n_{T_{max},Cu}} = \frac{n_s(1 - s_{T_{max},Al})}{n_s(1 - s_{T_{max},Cu})} \Rightarrow \frac{n_{T_{max},Al}}{n_{T_{max},Cu}} = \frac{1 - s_{T_{max},Al}}{1 - s_{T_{max},Cu}} \tag{6}$$

Solving (5) and (6):

$$\Rightarrow \frac{n_{T_{max},Al}}{n_{T_{max},Cu}} = \frac{1 - 2s_{T_{max},Cu}}{1 - s_{T_{max},Cu}} \Rightarrow \frac{n_{T_{max},Al}}{n_{T_{max},Cu}} < 1$$

Choice (2) is the answer.

4.25. Based on the information given in the problem, we have:

$$p = 4 \tag{1}$$

$$f_s = 50 \, Hz \tag{2}$$

$$n_1 = 1450 \, rpm \tag{3}$$

$$R_{r,2} = 3R_{r,1} \tag{4}$$

The synchronous speed can be calculated based on the number of poles of rotor and frequency of stator as follows:

$$n_s = \frac{120f_s}{p} \tag{5}$$

$$\Rightarrow n_s = \frac{120 \times 50}{4} \Rightarrow n_s = 1500 \, rpm \tag{6}$$

The slip can be calculated as follows:

$$s_1 = \frac{n_s - n_1}{n_s} \tag{7}$$

$$\Rightarrow s_1 = \frac{1500 - 1450}{1500} \Rightarrow s_1 = 0.033 \tag{8}$$

As we know, at speed around synchronous speed (stable operating region), the mechanical torque can be approximated by the following relation:

$$T_m \sim \frac{V^2}{f_s} \frac{s}{R_r'} \tag{9}$$

Since the power supply of motor is the same as the previous one, its voltage and frequency do not change. Moreover, the load of motor is constant; therefore, the torque will be the same. Thus, from Eq. (9), it is concluded that slip is proportional to rotor resistance. In other words:

$$s \propto R_r' \Rightarrow s \propto R_r \tag{10}$$

Solving (4) and (10):

$$s_2 = 3s_1 = 0.1 \tag{11}$$

The new speed of motor can be calculated as follows:

$$n_2 = n_s(1 - s_2) \tag{12}$$

$$\Rightarrow n_2 = 1500(1 - 0.1) \Rightarrow n_2 = 1350 \; rpm$$

Choice (3) is the answer.

4.26. Based on the information given in the problem, we have:

$$p = 6 \tag{1}$$

$$f_s = 50 \; Hz \tag{2}$$

$$n = 950 \; rpm \tag{3}$$

$$T_{load,1} = T_{FL} \tag{4}$$

$$T_{load,2} = 0.5 T_{FL} \tag{5}$$

The synchronous speed can be calculated based on the number of poles of rotor and frequency of stator as follows:

$$n_s = \frac{120 f_s}{p} \tag{6}$$

$$\Rightarrow n_s = \frac{120 \times 50}{6} \Rightarrow n_s = 1000 \; rpm \tag{7}$$

The slip can be calculated as follows:

$$s_1 = \frac{n_s - n_1}{n_s} \tag{8}$$

$$\Rightarrow s_1 = \frac{1000 - 950}{1000} \Rightarrow s_1 = 0.05 \tag{9}$$

As we know, at speed around synchronous speed (stable operating region), the mechanical torque can be approximated by the following relation:

$$T_m \sim \frac{V^2}{f_s} \frac{s}{R_r'} \tag{9}$$

Since the power supply of motor is the same as the previous one, its voltage and frequency do not change. In addition, the rotor resistance is constant. Hence, from Eq. (9), it is concluded that torque is proportional to slip. In other words:

$$T_e \propto s \tag{10}$$

Solving (4), (5), (9), and (10):

$$s_2 = 0.025 \tag{11}$$

The speed of motor at half full-load condition can be calculated as follows:

$$n_2 = n_s(1 - s_2) \tag{12}$$

$$\Rightarrow n_2 = 1000(1 - 0.025) \Rightarrow n_2 = 975 \ rpm$$

Choice (1) is the answer.

4.27. Based on the information given in the problem, we have:

$$p = 6 \tag{1}$$

$$f_s = 50 \ Hz \tag{2}$$

$$T_{out} = 120 \ N.m. \tag{3}$$

$$f_r = 2 \ Hz \tag{4}$$

$$P_{rot} = 2 \ kW \tag{5}$$

$$P_{cu,s} + P_c = 0.5 \ kW \tag{6}$$

The slip can be calculated as follows:

$$s = \frac{f_r}{f_s} \tag{7}$$

$$\Rightarrow s = \frac{2}{50} \Rightarrow s = 0.04 \tag{8}$$

The synchronous speed can be calculated based on the number of poles of rotor and frequency of stator as follows:

$$n_s = \frac{120 f_s}{p} \tag{9}$$

$$\Rightarrow n_s = \frac{120 \times 50}{6} \Rightarrow n_s = 1000 \ rpm \tag{10}$$

The speed of rotor can be determined as follows:

$$n = n_s(1 - s) \tag{11}$$

$$\Rightarrow n = 1000(1 - 0.04) \Rightarrow n = 960 \ rpm \tag{12}$$

The output power can be calculated as follows:

$$P_{out} = T_{out}\omega = T_{out}n \times \frac{2\pi}{60} \tag{13}$$

$$\Rightarrow P_{out} = 120 \times 960 \times \frac{2\pi}{60} \Rightarrow P_{out} = 12063.7 \ W \tag{14}$$

Figure 4.4 shows the power flow diagram of an induction motor. The developed electromagnetic power can be calculated as follows:

$$P_e = P_{out} + P_{rot} \tag{15}$$

$$\Rightarrow P_e = 12063.7 + 2000 = 14063.7 \ W \tag{16}$$

As we know, the air gap power can be calculated as follows:

$$P_{ag} = \frac{P_e}{1 - s} \tag{17}$$

$$\rightarrow P_{ag} - \frac{14063.7}{1 - 0.04} = 14649.7 \ W \tag{18}$$

Based on the power flow diagram, the input power can be calculated as follows:

$$P_{in} = P_{ag} + P_{cu,s} + P_c \tag{19}$$

$$\Rightarrow P_{in} = 14649.7 + 500 = 15146.7 \ W \tag{20}$$

Now, the efficiency of motor can be calculated as follows:

$$\eta = \frac{P_{out}}{P_{in}} \times 100 \tag{21}$$

$$\eta = \frac{12063.7}{15146.7} \times 100 \Rightarrow \eta = 79.6\%$$

Choice (3) is the answer.

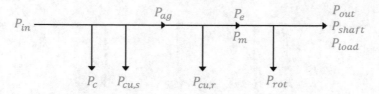

Fig. 4.4 The circuit of solution of Problem 4.27

4.28. Based on the information given in the problem, we have:

$$p = 4 \tag{1}$$

$$f_s = 50 \ Hz \tag{2}$$

$$n = 1440 \ rpm \tag{3}$$

$$P_{out} = 20 \ hp \tag{4}$$

The synchronous speed can be calculated based on the number of poles of rotor and frequency of stator as follows:

$$n_s = \frac{120 f_s}{p} \tag{5}$$

$$\Rightarrow n_s = \frac{120 \times 50}{4} \Rightarrow n_s = 1500 \ rpm \tag{6}$$

The slip can be calculated as follows:

$$s = \frac{n_s - n}{n_s} \tag{7}$$

$$\Rightarrow s = \frac{1500 - 1440}{1500} \Rightarrow s = 0.04 \tag{8}$$

Figure 4.5 shows the power flow diagram of an induction motor. Since the rotational power loss is not given, we can ignore it. Therefore, the developed electromagnetic power is as follows:

$$P_e = P_{out} = 20 \times 746 = 14920 \ W \tag{9}$$

As we know, the air gap power can be calculated as follows:

$$P_{ag} = \frac{P_e}{1 - s} \tag{10}$$

$$\Rightarrow P_{ag} = \frac{14920}{1 - 0.04} = 15541.66 \ W \tag{11}$$

In addition, the copper power loss of rotor can be calculated as follows:

$$P_{cu,r} = s P_{ag} \tag{12}$$

$$\Rightarrow P_{cu,r} = 0.04 \times 15541.66 \Rightarrow P_{cu,r} = 621.66 \ W$$

Choice (4) is the answer.

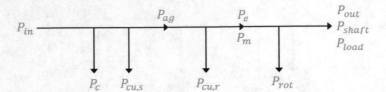

Fig. 4.5 The circuit of solution of problem 4.28

4.29. Based on the information given in the problem, we have:

$$p = 6 \tag{1}$$

$$f_s = 50 \, Hz \tag{2}$$

$$P_{out} = 12 \, hp \tag{3}$$

$$n = 950 \, rpm \tag{4}$$

$$P_{cu,s} = 1 \, kW \tag{5}$$

$$P_c = 0.5 \, kW \tag{6}$$

$$P_{rot} \approx 0 \tag{7}$$

The synchronous speed can be calculated based on the number of poles of rotor and frequency of stator as follows:

$$n_s = \frac{120 f_s}{p} \tag{8}$$

$$\Rightarrow n_s = \frac{120 \times 50}{6} \Rightarrow n_s = 1000 \, rpm \tag{9}$$

The slip can be calculated as follows:

$$s = \frac{n_s - n}{n_s} \tag{10}$$

$$\Rightarrow s = \frac{1000 - 950}{1000} \Rightarrow s = 0.05 \tag{11}$$

Figure 4.6 shows the power flow diagram of an induction motor. The rotational power loss is zero; thus, the developed electromagnetic power is as follows:

$$P_e = P_{out} + P_{rot} = 12 \times 746 + 0 = 8952 \, W \tag{12}$$

As we know, the copper power loss of rotor can be calculated as follows:

$$P_{cu,r} = sP_{ag} = s \times \frac{P_e}{1-s} = \frac{s}{1-s} P_e \tag{13}$$

$$\Rightarrow P_{cu,r} = \frac{0.05}{1-0.05} \times 8952 \Rightarrow P_{cu,r} = 471.15 \, W \tag{14}$$

The total power loss of motor can be calculated as follows:

$$P_{loss} = P_{cu,s} + P_c + P_{cu,r} + P_{rot} \tag{15}$$

$$\Rightarrow P_{loss} = 1000 + 500 + 471.15 + 0 = 1971.15 \ W \tag{16}$$

Now, the efficiency of motor can be calculated as follows:

$$\eta = \frac{P_{out}}{P_{out} + P_{loss}} \times 100 \tag{17}$$

$$\eta = \frac{8952}{8952 + 1971.15} \times 100 \Rightarrow \eta = 81.95 \approx 82\%$$

Choice (2) is the answer.

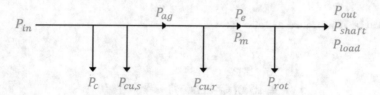

Fig. 4.6 The circuit of solution of Problem 4.29

4.30. Based on the information given in the problem, we have:

$$p = 4 \tag{1}$$

$$f_s = 50 \ Hz \tag{2}$$

$$\eta = 85\% \tag{3}$$

$$P_{out} = 25.5 \ kW \tag{4}$$

$$P_{cu,s} = P_{cu,r} = P_c = P_{rot} \tag{5}$$

As we know, efficiency can be calculated as follows:

$$\eta = \frac{P_{out}}{P_{out} + P_{loss}} \times 100 \tag{6}$$

$$85 = \frac{25.5}{25.5 + P_{loss}} \times 100 \Rightarrow P_{loss} = 4.5 \ kW \tag{7}$$

Solving (5) and (7):

$$P_{cu,s} = P_{cu,r} = P_c = P_{rot} = \frac{4.5}{4} = 1.125 \ kW \tag{8}$$

Figure 4.7 shows the power flow diagram of an induction motor. The developed electromagnetic power can be calculated as follows:

$$P_e = P_{out} + P_{rot} = 25.5 + 1.125 = 26.625\ W \tag{9}$$

As we know, the relation between copper power loss of rotor and electromagnetic power is as follows:

$$P_{cu,r} = \frac{s}{1-s} P_e \tag{10}$$

$$\Rightarrow 1.125 = \frac{s}{1-s} \times 25.625 \Rightarrow s = 0.0405 \tag{11}$$

The synchronous speed can be calculated based on the number of poles of rotor and frequency of stator as follows:

$$n_s = \frac{120 f_s}{p} \tag{12}$$

$$\Rightarrow n_s = \frac{120 \times 50}{4} \Rightarrow n_s = 1500\ rpm \tag{13}$$

The speed of rotor can be determined as follows:

$$n = n_s(1 - s) \tag{14}$$

$$\Rightarrow n = 1500(1 - 0.0405) \Rightarrow n = 1440\ rpm$$

Choice (3) is the answer.

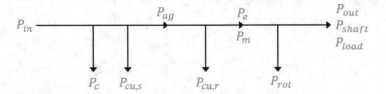

Fig. 4.7 The circuit of solution of Problem 4.30

4.31. Based on the information given in the problem, we have:

$$p = 8 \tag{1}$$

$$f_s = 50\ Hz \tag{2}$$

$$n = 700\ rpm \tag{3}$$

$$P_{in} = 5\ kW \tag{4}$$

$$P_{cu,s} + P_c = 100\ W \tag{5}$$

$$P_{rot} = 150\ W \tag{6}$$

The synchronous speed can be calculated based on the number of poles of rotor and frequency of stator as follows:

$$n_s = \frac{120 f_s}{p} \tag{7}$$

$$\Rightarrow n_s = \frac{120 \times 50}{8} \Rightarrow n_s = 750 \ rpm \tag{8}$$

The slip can be calculated as follows:

$$s = \frac{n_s - n}{n_s} \tag{9}$$

$$\Rightarrow s = \frac{750 - 700}{750} \Rightarrow s = 0.066 \tag{10}$$

Figure 4.8 shows the power flow diagram of an induction motor. The air gap power can be calculated as follows:

$$P_{ag} = P_{in} - (P_{cu,s} + P_c) = 5000 - 100 = 4900 \tag{11}$$

As we know, the developed electromagnetic power can be calculated as follows:

$$P_e = (1 - s)P_{ag} = (1 - 0.066) \times 4900 = 4570 \ W \tag{12}$$

Based on the power flow diagram, the output power can be calculated as follows:

$$P_{out} = P_e - P_{rot} = 4570 - 150 = 4420 \ W \tag{13}$$

The efficiency of motor can be calculated as follows:

$$\eta = \frac{P_{out}}{P_{in}} \times 100 \tag{14}$$

$$\Rightarrow \eta = \frac{4420}{5000} \times 100 \Rightarrow \eta = 88.4\%$$

Choice (4) is the answer.

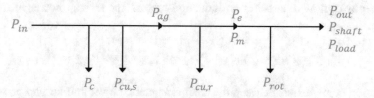

Fig. 4.8 The circuit of solution of Problem 4.31

4.32. Based on the information given in the problem, we have:

$$p_1 = 2 \tag{1}$$

$$p_2 = 4 \tag{2}$$

$$f_s = 50 \ Hz \tag{3}$$

$$s_1 = 0.08 \tag{4}$$

$$P_{cu,s} + P_c \approx 0 \ W \tag{5}$$

The primary synchronous speed can be calculated based on the number of poles of rotor and frequency of stator as follows:

$$n_{s,1} = \frac{120 f_s}{p_1} \tag{6}$$

$$\Rightarrow n_{s,1} = \frac{120 \times 50}{2} \Rightarrow n_{s,1} = 3000 \ rpm \tag{7}$$

Thus, the corresponding primary speed of rotor is as follows:

$$n_1 = n_{s,1}(1 - s_1) \tag{8}$$

$$\Rightarrow n_1 = 3000(1 - 0.08) \Rightarrow n_1 = 2760 \ rpm \tag{9}$$

When the number of poles of motor is changed from 2 to 4, its speed will temporarily remain constant; however, the synchronous speed is updated. Therefore:

$$\Rightarrow n_2 = 2760 \ rpm \tag{10}$$

$$n_{s,2} = \frac{120 f_s}{p_3} \Rightarrow n_{s,2} = \frac{120 \times 50}{4} \Rightarrow n_{s,2} = 1500 \ rpm \tag{11}$$

The new slip can be calculated as follows:

$$s_2 = \frac{n_{s,2} - n_2}{n_{s,2}} \tag{12}$$

$$\Rightarrow s_2 = \frac{1500 - 2760}{1500} \Rightarrow s_2 = -0.84 \tag{13}$$

As can be noticed from the new slip, the machine temporarily changes to a generator, since $s_2 < 0$.

Figure 4.9 shows the power flow diagram of an induction motor. The air gap power in the second condition can be calculated as follows:

$$P_{ag,2} = P_{in,2} - (P_{cu,s,2} + P_{c,2}) = P_{in,2} - 0 \Rightarrow P_{ag,2} = P_{in,2} \tag{14}$$

As we know, the relation below exists between the electromagnetic power and air gap power in the second condition:

$$P_{e,2} = (1 - s_2) P_{ag,2} \tag{15}$$

Solving (14) and (15):

$$P_{e,2} = (1 - s_2) P_{in,2} \Rightarrow \frac{P_{e,2}}{P_{in,2}} = 1 - s_2 = 1 - (-0.84) \tag{16}$$

$$\Rightarrow \frac{P_{e,2}}{P_{in,2}} = 1.84$$

Choice (2) is the answer.

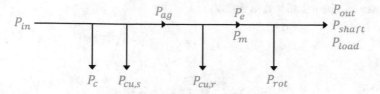

Fig. 4.9 The circuit of solution of Problem 4.32

4.33. Based on the information given in the problem, we have:

$$p = 6 \tag{1}$$

$$f_s = 50 \, Hz \tag{2}$$

$$P_{out} = 50 \, kW \tag{3}$$

$$n = 970 \, rpm \tag{4}$$

$$P_c = 1 \, kW \tag{5}$$

$$P_{cu,s} = 1 \, kW \tag{6}$$

$$P_{rot} = 1.5 \, kW \tag{7}$$

The synchronous speed can be calculated based on the number of poles of rotor and frequency of stator as follows:

$$n_s = \frac{120 f_s}{p} \tag{8}$$

$$\Rightarrow n_s = \frac{120 \times 50}{6} \Rightarrow n_s = 1000 \, rpm \tag{9}$$

The slip can be calculated as follows:

$$s = \frac{n_s - n}{n_s} \tag{9}$$

$$\Rightarrow s = \frac{1000 - 970}{1000} \Rightarrow s = 0.03 \tag{10}$$

Figure 4.10 shows the power flow diagram of an induction motor. The developed electromagnetic power can be calculated as follows:

$$P_e = P_{out} + P_{rot} = 50 + 1.5 = 51.5 \, kW \tag{11}$$

As we know, the air gap power can be calculated as follows:

$$P_{ag} = \frac{1}{1-s} P_e = \frac{1}{1 - 0.03} \times 51.5 = 53 \, kW \tag{12}$$

Based on the power flow diagram, the input power can be calculated as follows:

$$P_{in} = P_{ag} + P_{cu,s} + P_c = 53 + 1 + 1 = 55 \, kW \tag{13}$$

The efficiency of motor can be calculated as follows:

$$\eta = \frac{P_{out}}{P_{in}} \times 100 \tag{14}$$

$$\Rightarrow \eta = \frac{50}{55} \times 100 \Rightarrow \eta = 91\%$$

Choice (1) is the answer.

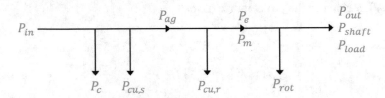

Fig. 4.10 The circuit of solution of Problem 4.33

4.34. Based on the information given in the problem, we have:

$$p = 4 \tag{1}$$

$$P_{out} = 9.5 \; kW \tag{2}$$

$$n = 1455 \; rpm \tag{3}$$

$$f_s = 50 \; Hz \tag{4}$$

$$P_{rot} = 0.2 \; kW \tag{5}$$

$$I_r = 5 \; A \tag{6}$$

The synchronous speed can be calculated based on the number of poles of rotor and frequency of stator as follows:

$$n_s = \frac{120 f_s}{p} \tag{7}$$

$$\Rightarrow n_s = \frac{120 \times 50}{4} \Rightarrow n_s = 1500 \; rpm \tag{8}$$

The slip can be calculated as follows:

$$s = \frac{n_s - n}{n_s} \tag{9}$$

$$\Rightarrow s = \frac{1500 - 1455}{1500} \Rightarrow s = 0.03 \tag{10}$$

As we know, the electromagnetic power can be calculated as follows:

$$P_e = P_{out} + P_{rot} \tag{11}$$

$$\Rightarrow P_e = 9.5 + 0.2 = 9.7 \; kW \tag{12}$$

Moreover, the copper power loss of rotor can be calculated as follows:

$$P_{cu,r} = \frac{s}{1-s}P_e = \frac{0.03}{1-0.03} \times 9.7 = 0.3\ kW = 300\ W \tag{13}$$

On the other hand, the copper power loss of rotor of a three-phase induction motor can be directly calculated as follows:

$$P_{cu,r} = 3R_r(I_r)^2 \tag{14}$$

$$\Rightarrow P_{cu,r} = 3R_r(5)^2 = 75R_r \tag{15}$$

Solving (13) and (15):

$$R_r = \frac{300}{75} \Rightarrow R_r = 4\ \Omega/phase$$

Choice (4) is the answer.

4.35. Based on the information given in the problem, we have:

$$p = 4 \tag{1}$$

$$f_s = 60\ Hz \tag{2}$$

$$T_{load} = \text{Const.} \tag{3}$$

$$n = 1710\ rpm \tag{4}$$

$$\varphi_{ag} = \text{Const.} \tag{5}$$

$$P_{rot} = \text{Const.} \tag{6}$$

$$R_{r,2} = 5R_{r,1} \tag{7}$$

The synchronous speed can be calculated based on the number of poles of rotor and frequency of stator as follows:

$$n_s = \frac{120f_s}{p} \tag{8}$$

$$\Rightarrow n_s = \frac{120 \times 60}{4} \Rightarrow n_s = 1800\ rpm \tag{9}$$

The slip can be calculated as follows:

$$s_1 = \frac{n_s - n}{n_s} \tag{10}$$

$$\Rightarrow s_1 = \frac{1800 - 1710}{1800} \Rightarrow s_1 = 0.05 \tag{11}$$

As we know, at speed around synchronous speed (stable operating region), the mechanical torque can be approximated by the following relation:

$$T_m \sim \frac{V^2}{f_s}\frac{s}{R_r'} \tag{12}$$

Since the air gap magnetic flux is constant, the voltage and frequency of motor remain constant. Therefore:

$$\frac{T_{m,1}}{T_{m,2}} = \frac{s_1}{s_2} \times \frac{R'_{r,2}}{R'_{r,1}} \tag{13}$$

The load torque and rotational power loss are constant. Therefore, the mechanical torque will be unchanged:

$$T_{m,2} = T_{m,1} \tag{14}$$

Solving (7), (11), (13), and (14):

$$1 = \frac{s_1}{s_2} \times 5 \Rightarrow s_2 = 5s_1 = 5 \times 0.05 = 0.25 \tag{15}$$

As we know, the relation between developed electromagnetic power and air gap power is as follows:

$$\frac{P_e}{P_{ag}} = 1 - s \tag{16}$$

Therefore, for the second condition, we have:

$$\frac{P_{e,2}}{P_{ag,2}} = 1 - s_2 = 1 - 0.25 \Rightarrow \frac{P_{e,2}}{P_{ag,2}} = 0.75$$

Choice (2) is the answer.

4.36. Based on the information given in the problem, we have:

$$p = 4 \tag{1}$$

$$f_s = 50 \ Hz \tag{2}$$

$$T_{st} = 1.5 T_{FL} \tag{3}$$

$$T_{max} = 3 T_{FL} \tag{4}$$

As we know, the relation between starting torque and maximum torque is as follows:

$$\frac{T_{st}}{T_{max}} = \frac{2 s_{T_{max}}}{\left(s_{T_{max}}\right)^2 + 1} \tag{5}$$

$$\Rightarrow \frac{1.5}{3} = \frac{2 s_{T_{max}}}{\left(s_{T_{max}}\right)^2 + 1} \Rightarrow s_{T_{max}} = 0.2679 \tag{6}$$

Moreover, the relation between rated torque and maximum torque is as follows:

$$\frac{T_{FL}}{T_{max}} = \frac{2 s_{T_{max}} s_{T_{FL}}}{\left(s_{T_{max}}\right)^2 + \left(s_{T_{FL}}\right)^2} \tag{7}$$

$$\Rightarrow \frac{1}{3} = \frac{2 \times 0.2679 \times s_{T_{FL}}}{\left(0.2679\right)^2 + \left(s_{T_{FL}}\right)^2} \Rightarrow s_{T_{FL}} = 0.04597 \tag{8}$$

The synchronous speed can be calculated based on the number of poles of rotor and frequency of stator as follows:

$$n_s = \frac{120 f_s}{p} \tag{9}$$

$$\Rightarrow n_s = \frac{120 \times 50}{4} \Rightarrow n_s = 1500 \ rpm \tag{10}$$

Now, the speed of motor at full-load condition can be calculated as follows:

$$n_{FL} = n_s(1 - s_{FL}) \tag{11}$$

$$\Rightarrow n_{FL} = 1500(1 - 0.04597) \Rightarrow n_{FL} = 1431 \ rpm$$

Choice (1) is the answer.

4.37. Based on the information given in the problem, we have:

$$f_{s,1} = 50 \ Hz \tag{1}$$

$$n_1 = 1425 \ rpm \tag{2}$$

$$\frac{V}{f} = \text{Const.} \tag{3}$$

$$T_{load} = \text{Const.} \tag{4}$$

$$V_2 = 1.5 V_1 \tag{5}$$

$$f_{s,2} = 1.5 f_{s,1} \tag{6}$$

$$Z_s \approx 0 \ \Omega/phase \tag{7}$$

Based on the speed of motor given in the problem, we can determine the synchronous speed of machine as follows:

$$n_s = \frac{120 f_s}{p} \tag{8}$$

$$\Rightarrow n_{s,1} = \frac{120 \times 50}{p} = \frac{6000}{p} \Rightarrow \begin{cases} p = 1 \Rightarrow n_{s,1} = 6000 \\ p = 2 \Rightarrow n_{s,1} = 3000 \\ p = 4 \Rightarrow n_{s,1} = 1500 \\ p = 6 \Rightarrow n_{s,1} = 1000 \\ p = 8 \Rightarrow n_{s,1} = 7500 \end{cases} \tag{9}$$

Since speed of an induction machine is close to its synchronous speed during stable operation (rated condition), it is clear that $n_{s,\ 1} = 1500$ for this machine because $n_1 = 1425 \ rpm$.

Thus, the primary slip of motor can be calculated as follows:

$$s_1 = \frac{n_{s,1} - n_1}{n_s} \tag{10}$$

$$\Rightarrow s_1 = \frac{1500 - 1425}{1500} \Rightarrow s_1 = 0.05 \tag{11}$$

As we know, at speed around synchronous speed (stable operating region), the mechanical torque can be approximated by the following relation:

$$T_m \sim \frac{V^2}{f_s}\frac{s}{R'_r} \tag{12}$$

Herein, since the resistance of rotor and load torque (and consequently mechanical torque) are constant, we can write:

$$1 = \left(\frac{V_2}{V_1}\right)^2 \left(\frac{f_{s,1}}{f_{s,2}}\right)\left(\frac{s_2}{s_1}\right) \times 1 \tag{13}$$

$$\Rightarrow 1 = (1.5)^2 \left(\frac{1}{1.5}\right)\left(\frac{s_2}{0.05}\right) \Rightarrow s_2 = \frac{1}{30} \tag{14}$$

The new synchronous speed of machine can be calculated as follows:

$$n_{s,2} = \frac{120 f_{s,2}}{p} \tag{15}$$

$$\Rightarrow n_{s,2} = \frac{120 \times (1.5 f_{s,1})}{p} = 1.5 \times \frac{120 f_{s,1}}{p} = 1.5 n_{s,1} = 1.5 \times 1500 = 2250 \; rpm \tag{16}$$

Then, the new speed of motor can be calculated as follows:

$$n_2 = n_{s,2}(1 - s_2) \tag{17}$$

$$\Rightarrow n_2 = 2250\left(1 - \frac{1}{30}\right) \Rightarrow n_2 = 2175 \; rpm$$

Choice (2) is the answer.

Problems: Synchronous Machines

5

Abstract

In this chapter, the basic and advanced problems related to the three-phase synchronous motors and generators are solved. The subjects include the determination of excitation voltage, power angle, steady-state stability limit, armature current, active power, reactive power, and power factor of a synchronous machine using its equivalent circuit at different load and stability conditions. In this chapter, the problems are categorized in different levels based on their difficulty levels (easy, normal, and hard) and calculation amounts (small, normal, and large); however, the problems are not ordered from the easiest problems to the most difficult ones because some of the problems are related to the previous ones.

5.1. A three-phase, 5 kVA, 208 V, four-pole, 60 Hz, star-connected synchronous generator is delivering the rated apparent power at the power factor of 0.8 lagging. Determine the excitation voltage and power angle if the generator has negligible stator winding resistance and a synchronous reactance of 8 ohms per phase.

Difficulty level ○ Easy ● Normal ○ Hard
Calculation amount ○ Small ● Normal ○ Large

1) $E_f = 206.9$ V/phase, $\delta = 25.5°$
2) $E_f = 120$ V/phase, $\delta = 0°$
3) $E_f = 206.9$ V/phase, $\delta = -25.5°$
4) $E_f = 120$ V/phase, $\delta = -25.5°$

5.2. In problem 5.1, calculate the theoretical steady-state (static) stability limit of generator.

Difficulty level ○ Easy ● Normal ○ Hard
Calculation amount ● Small ○ Normal ○ Large

1) 4.56 kW
2) 18.64 kW
3) 0.932 kW
4) 9.32 kW

5.3. In problem 5.1, calculate the stator (armature) current and power factor when the generator is delivering the maximum power based on its theoretical static stability limit.

Difficulty level ○ Easy ● Normal ○ Hard
Calculation amount ○ Small ● Normal ○ Large

1) $I_a = 15.95 \angle 30.1°$ A, 0.865 lagging
2) $I_a = 29.9 \angle 30.1°$ A, 0.865 leading
3) $I_a = 29.9 \angle 30.1°$ A, 0
4) $I_a = 15.95 \angle 30.1°$ A, 0.865 leading

5.4. In Problem 5.1, by fixing the input power of generator (prime mover power), the field excitation current is increased about 20%. Calculate the stator current, power factor, and reactive power supplied by the generator.

Difficulty level   ○ Easy   ○ Normal   ● Hard
Calculation amount   ○ Small   ○ Normal   ● Large

1) $I_a = 17.86\angle 51.5° \, A$, 0.62 leading, 5.03 $kVAr$
2) $I_a = 17.86\angle -51.5° \, A$, 0.62 leading, 10.06 $kVAr$
3) $I_a = 17.86\angle -51.5° \, A$, 0.62 lagging, 5.03 $kVAr$
4) $I_a = 17.86\angle 51.5° \, A$, 0.62 lagging, 10.06 $kVAr$

5.5. A three-phase, four-pole, 60 Hz, 208 V, star-connected synchronous motor is supplied by a three-phase, 208 V, 60 Hz power source. The field excitation is adjusted to have the unity power factor when the motor is consuming 3 kW power. Determine the excitation voltage and power angle. Herein, assume that the motor has negligible armature winding resistance and a synchronous reactance of 8 ohms per phase.

Difficulty level   ○ Easy   ● Normal   ○ Hard
Calculation amount   ○ Small   ● Normal   ○ Large

1) $E_f = 137.35 \, V/phase$, $\delta = 29°$
2) $E_f = 137.35 \, V/phase$, $\delta = -29°$
3) $E_f = 208 \, V/phase$, $\delta = -29°$
4) $E_f = 208 \, V/phase$, $\delta = 29°$

5.6. In Problem 5.5, calculate the theoretical maximum torque developed by the motor if the field excitation is held constant and the shaft load is slowly increased.

Difficulty level   ○ Easy   ● Normal   ○ Hard
Calculation amount   ● Small   ○ Normal   ○ Large

1) 32.8 $N. \, m$
2) 6280 $N. \, m$
3) 65.6 $N. \, m$
4) 16.4 $N. \, m$

5.7. A three-phase, 11 kV, 5 MVA, 60 Hz synchronous machine is connected to a 11 kV, 60 Hz bus and operated as a synchronous condenser. Calculate the power angle and stator current. Herein, assume that $R_s = 0$ and $X_s = 10 \, \Omega/phase$. In addition, assume a normal excitation for the machine.

Difficulty level   ○ Easy   ● Normal   ○ Hard
Calculation amount   ○ Small   ● Normal   ○ Large

1) 90 °, $635.1\angle 0° \, A$
2) 90 °, $1\angle 90° \, A$
3) 0 °, $1\angle -90° \, A$
4) 0 °, 0 A

5.8. In Problem 5.7, calculate the stator current and power factor if the excitation is increased to 150% of the normal excitation.

Difficulty level   ○ Easy   ● Normal   ○ Hard
Calculation amount   ○ Small   ● Normal   ○ Large

1) $317.55\angle 90° \, A$, 0 leading
2) $317.55\angle -90° \, A$, 0 lagging
3) $317.55\angle -90° \, A$, 0 leading
4) 31.755 A, 0 lagging

5.9. In Problem 5.7, calculate the stator current and power factor if the excitation is decreased to 50% of the normal excitation.

Difficulty level   ○ Easy   ● Normal   ○ Hard
Calculation amount   ○ Small   ● Normal   ○ Large

1) 31.755 A, 0 lagging
2) $317.55\angle 90° \, A$, 0 leading
3) $317.55\angle -90° \, A$, 0 leading
4) $317.55\angle -90° \, A$, 0 lagging

5.10. A three-phase, 50 MVA, 30 kV, 60 Hz synchronous generator has the armature winding resistance and synchronous reactance of 0 and 9 Ω/phase, respectively. The generator is delivering the rated power at 0.8 lagging power factor and rated voltage to the infinite bus. Determine the excitation voltage and power angle of generator.

Difficulty level   ○ Easy   ● Normal   ○ Hard
Calculation amount   ○ Small   ● Normal   ○ Large

1) $E_f = 23558$ *V/phase*, $\delta = 17.1°$
2) $E_f = 17320$ *V/phase*, $\delta = -17.1°$
3) $E_f = 23558$ *V/phase*, $\delta = -17.1°$
4) $E_f = 17320$ *V/phase*, $\delta = 17.1°$

5.11. In Problem 5.10, if the excitation is kept constant, what is the steady-state maximum power that the generator can deliver without losing its synchronism? In addition, calculate the stator (armature) current in this condition.

Difficulty level   ○ Easy   ● Normal   ○ Hard
Calculation amount   ○ Small   ● Normal   ○ Large

1) 45.3 MW, $I_a = 324.88\angle 36.32°$ A
2) 45.3 MW, $I_a = 3248.8\angle 36.32°$ A
3) 136 MW, $I_a = 3248.8\angle 36.32°$ A
4) 136 MW, $I_a = 324.88\angle 30.10°$ A

5.12. In Problem 5.10, the excitation is kept constant, but the driving torque is reduced until the generator delivers 25 MW. Calculate the armature current and power factor of machine in this new condition.

Difficulty level   ○ Easy   ● Normal   ○ Hard
Calculation amount   ○ Small   ● Normal   ○ Large

1) $I_a = 807.4\angle -53.43°$ A, 0.596 lagging
2) $I_a = 807.4\angle 53.43°$ A, 0.596 leading
3) $I_a = 80.74\angle -53.43°$ A, 0.596 lagging
4) $I_a = 80.74\angle 53.43°$ A, 0.596 lagging

Abstract

In this chapter, the problems of the fifth chapter are fully solved, in detail, step-by-step, and with different methods.

6.1. Based on the information given in the problem, the stator winding of synchronous generator is star-connected. Moreover, we have:

$$S = 5 \, kVA \tag{1}$$

$$V_t = 208 \, V \tag{2}$$

$$p = 4 \tag{3}$$

$$f_s = 60 \, Hz \tag{4}$$

$$PF = 0.8 \text{ Lagging} \tag{5}$$

$$R_s \approx 0 \, \Omega/phase \tag{6}$$

$$X_s = 8 \, \Omega/phase \tag{7}$$

Since the synchronous generator is a star-connected machine, its rated phase voltage is as follows:

$$V_{t,ph} = \frac{208}{\sqrt{3}} = 120 \, V \tag{8}$$

Normally, the terminal voltage of synchronous generator is assumed as the reference voltage. Therefore:

$$\mathbf{V_{t,ph}} = 120\angle 0\,° \, V \tag{9}$$

The rated current of machine can be calculated as follows:

$$I_t = \frac{S}{\sqrt{3}V_t} = \frac{5000}{\sqrt{3} \times 208} = 13.9 \, A \tag{10}$$

Solving (5) and (10):

$$\mathbf{I_t} = I_t\angle - \cos^{-1}PF \tag{11}$$

$$\Rightarrow \mathbf{I_t} = 13.9\angle - \cos^{-1}0.8 = 13.9\angle - 36.9° \tag{12}$$

In (12), negative sign is applied as the power factor of load is lagging.

Figure 6.1 shows the per-phase equivalent circuit of a synchronous generator. By applying KVL in the loop, we have:

$$\mathbf{E_{f,ph}} = \mathbf{V_{t,ph}} + (R_s + jX_s)\mathbf{I_t} \tag{13}$$

$$\Rightarrow \mathbf{E_{f,ph}} = 120\angle 0° + (0 + j8)(13.9\angle - 36.9°) = 206.9\angle 25.5° \tag{14}$$

$$\Rightarrow \begin{cases} E_{f,ph} = 206.9\ V \\ \delta = 25.5° \end{cases}$$

Choice (1) is the answer.

The value of phase angle of excitation voltage (power angle) is positive because the machine is operated as a generator.

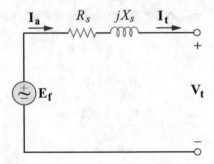

Fig. 6.1. The circuit of solution of Problem 6.1

6.2. Based on the information given in the problem, we have:

$$V_{t,ph} = 120\ V \tag{1}$$

$$E_{f,ph} = 206.9\ V \tag{2}$$

$$X_s = 8\ \Omega/phase \tag{3}$$

The theoretical steady-state (static) stability limit of a three-phase generator or the theoretical maximum deliverable power of a three-phase generator can be calculated as follows:

$$P_{max} = \frac{3V_{t,ph}E_{f,ph}}{X_s} \tag{4}$$

$$\Rightarrow P_{max} = \frac{3 \times 206.9 \times 120}{8} \Rightarrow P_{max} = 9.32\ kW$$

Choice (4) is the answer.

6.3. Based on the information given in the problem, we have:

$$\mathbf{V_{t,ph}} = 120\angle 0°\ V \tag{1}$$

$$E_{f,ph} = 206.9\ V \tag{2}$$

$$R_s \approx 0 \; \Omega/phase, \quad X_s = 8 \; \Omega/phase \tag{3}$$

The theoretical steady-state (static) stability limit of a generator happens when its power angle is 90°. In other words:

$$\delta_{max} = 90° \tag{4}$$

Figure 6.2 shows the per-phase equivalent circuit of a synchronous generator. By applying KVL in the loop, we have:

$$\mathbf{E}_{\mathbf{f,ph}} = \mathbf{V}_{\mathbf{t,ph}} + (R_s + jX_s)\mathbf{I_t} \tag{5}$$

$$\Rightarrow \mathbf{I_t} = \frac{\mathbf{E}_{\mathbf{f,ph}} - \mathbf{V}_{\mathbf{t,ph}}}{R_s + jX_s} = \frac{206.9\angle90° - 120\angle0°}{j8} \tag{6}$$

$$\Rightarrow \mathbf{I_a} = \mathbf{I_t} = 29.9\angle30.1°A$$

$$PF = \cos(\theta_V - \theta_I) = \cos(0 - 30.1°) \Rightarrow PF = 0.865$$

Since $(\theta_V - \theta_I) < 0$, the power factor is leading.

Choice (2) is the answer.

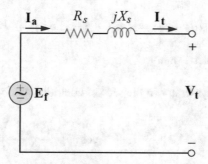

Fig. 6.2. The circuit of solution of Problem 6.3

6.4. Based on the information given in the problem, we have:

$$P_{prime\;mover,2} = P_{prime\;mover,1} \tag{1}$$

$$I_{f,2} = 1.2I_{f,1} \tag{2}$$

$$\mathbf{E}_{\mathbf{f,ph,1}} = 206.9\angle25.5° \tag{3}$$

$$\mathbf{V}_{\mathbf{t,ph}} = 120\angle0° \; V \tag{4}$$

$$R_s \approx 0 \; \Omega/phase, \quad X_s = 8 \; \Omega/phase \tag{5}$$

From (1), we can conclude that:

$$P_2 = P_1 \tag{6}$$

$$\Rightarrow \frac{3V_{t,ph}E_{f,ph,2}}{X_s}\sin\delta_2 = \frac{3V_{t,ph}E_{f,ph,1}}{X_s}\sin\delta_1 \tag{7}$$

$$\Rightarrow E_{f,ph,2}\sin\delta_2 = E_{f,ph,1}\sin\delta_1 \Rightarrow \sin\delta_2 = \frac{E_{f,ph,1}}{E_{f,ph,2}}\sin\delta_1 \tag{8}$$

From (2), we can conclude that:

$$E_{f,ph,2} = 1.2E_{f,ph,1} = 1.2 \times 206.9 = 248.28 \ V \tag{9}$$

Solving (8) and (9):

$$\Rightarrow \sin\delta_2 = \frac{1}{1.2}\sin 25.5° \Rightarrow \sin\delta_2 = 0.3587 \Rightarrow \delta_2 = 21° \tag{10}$$

Solving (9) and (10):

$$\mathbf{E_{f,ph,2}} = 248.28 \angle 21° \ V \tag{11}$$

Figure 6.3 shows the per-phase equivalent circuit of a synchronous generator. By applying KVL in the loop, we have:

$$\mathbf{E_{f,ph,2}} = \mathbf{V_{t,ph}} + (R_s + jX_s)\mathbf{I_{t,2}} \tag{12}$$

$$\Rightarrow \mathbf{I_{t,2}} = \frac{\mathbf{E_{f,ph,2}} - \mathbf{V_{t,ph}}}{R_s + jX_s} = \frac{248.28\angle 21° - 120\angle 0°}{j8} \tag{13}$$

$$\Rightarrow \mathbf{I_{a,2}} = \mathbf{I_{t,2}} = 17.86\angle -51.5° A \tag{14}$$

$$PF_2 = \cos(\theta_V - \theta_I) = \cos(0 - (-51.5°)) \Rightarrow PF_2 = 0.62 \tag{15}$$

Since $(\theta_V - \theta_I) > 0$, the power factor is lagging.

The reactive power delivered by the generator in the second condition can be calculated as follows:

$$Q_2 = 3\left(\frac{V_{t,ph}E_{f,ph,2}}{X_s}\cos\delta_2 - \frac{(V_{t,ph})^2}{X_s}\right) \tag{16}$$

$$\Rightarrow Q_2 = 3\left(\frac{120 \times 248.28}{8}\cos 21° - \frac{(120)^2}{8}\right) \Rightarrow Q_2 = 5.03 \ kVAr \tag{17}$$

Or we can use the following relation to calculate the reactive power supplied by the generator in the second condition:

$$Q_2 = 3V_{t,ph}I_{t,2}\sin(\theta_V - \theta_I) \tag{18}$$

$$\Rightarrow Q_2 = 3 \times 120 \times 17.86 \times \sin(51.5°) \Rightarrow Q_2 = 5.03 \ kVAr$$

Choice (3) is the answer.

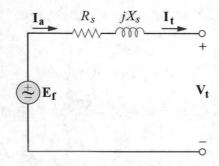

Fig. 6.3. The circuit of solution of Problem 6.4

6.5. Based on the information given in the problem, we have:

$$V_t = 208 \ V \tag{1}$$

$$PF = \cos\theta = 1 \tag{2}$$

$$P = 3 \ kW \tag{3}$$

$$R_s \approx 0 \ \Omega/phase \tag{4}$$

$$X_s = 8 \ \Omega/phase \tag{5}$$

Since the synchronous generator is a star-connected machine, its rated phase voltage is as follows:

$$V_{t,ph} = \frac{208}{\sqrt{3}} = 120 \ V \tag{6}$$

Normally, the terminal voltage of synchronous generator is assumed as the reference voltage. Therefore:

$$\mathbf{V_{t,ph}} = 120\angle 0^\circ \ V = 120 \ V \tag{7}$$

The rated current of machine can be calculated as follows:

$$I_t = \frac{P}{\sqrt{3}V_t \cos\theta} = \frac{3000}{\sqrt{3} \times 208 \times 1} = 8.33 \ A \tag{8}$$

Solving (2) and (8):

$$\mathbf{I_t} = I_t \angle \cos^{-1}(PF) \tag{9}$$

$$\Rightarrow \mathbf{I_t} = 8.33\angle \cos^{-1}(1) = 8.33\angle 0^\circ = 8.33 \ A \tag{10}$$

Figure 6.4 shows the per-phase equivalent circuit of a synchronous motor. By applying KVL in the loop, we have:

$$\mathbf{E_{f,ph}} = \mathbf{V_{t,ph}} - (R_s + jX_s)\mathbf{I_t} \tag{11}$$

$$\Rightarrow \mathbf{E_{f,ph}} = 120 - (0 + j8)(8.33) = 137.35\angle -29^\circ \ V \tag{12}$$

$$\Rightarrow \begin{cases} E_{f,ph} = 137.35 \ V \\ \delta = -29^\circ \end{cases}$$

Choice (2) is the answer.

The value of phase angle of excitation voltage (power angle) is negative because the machine is operated as a motor.

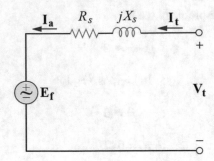

Fig. 6.4. The circuit of solution of Problem 6.5

6.6. Based on the information given in the problem, we have:

$$V_{t,ph} = 120 \ V \tag{1}$$

$$E_{f,ph} = 137.35 \ V \tag{2}$$

$$X_s = 8 \ \Omega/phase \tag{3}$$

$$f_s = 60 \ Hz \tag{4}$$

$$p = 4 \tag{5}$$

The theoretical steady-state (static) stability limit of a three-phase generator or the theoretical maximum deliverable power of a three-phase generator can be calculated as follows:

$$P_{max} = \frac{3V_{t,ph}E_{f,ph}}{X_s} \tag{6}$$

$$\Rightarrow P_{max} = \frac{3 \times 137.35 \times 120}{8} \Rightarrow P_{max} = 6185 \ W \tag{7}$$

The theoretical maximum torque can be calculated as follows:

$$T_{max} = \frac{P_{max}}{\omega_s} = \frac{P_{max}}{n_s \times \frac{2\pi}{60}} = \frac{P_{max}}{\frac{120f_s}{p} \times \frac{2\pi}{60}} \tag{8}$$

$$\Rightarrow T_{max} = \frac{6185}{\frac{120 \times 60}{4} \times \frac{2\pi}{60}} \Rightarrow T_{max} = 32.8 \ N.m.$$

Choice (1) is the answer.

6.7. Based on the information given in the problem, we have:

$$V_t = 11 \ k \ V \tag{1}$$

$$S = 5 \ MVA \tag{2}$$

$$R_s \approx 0 \ \Omega/phase \tag{3}$$

$$X_s = 8 \ \Omega/phase \tag{4}$$

Since the machine is operated as a synchronous condenser, it does not generate or consume any active power ($P = 0$). Therefore, based on the relation below, its power angle is zero:

$$P = \frac{3V_{t,ph}E_{f,ph}}{X_s} \sin\delta \Rightarrow 0 = \frac{3V_{t,ph}E_{f,ph}}{X_s} \sin\delta \Rightarrow \delta = 0° \tag{5}$$

Moreover, based on the relation below, the armature current is zero:

$$P = 3V_{t,ph}I_a \cos(\theta) \Rightarrow 0 = 3V_{t,ph}I_a \cos(\theta) \Rightarrow I_a = 0\ A \tag{6}$$

Choice (4) is the answer.

Figure 6.5 shows the per-phase equivalent circuit of a synchronous generator. Since $I_a = 0$, we have:

$$\mathbf{E_{f,ph}} = \mathbf{V_{t,ph}} \tag{7}$$

Since the synchronous generator is a star-connected machine, its rated phase voltage is as follows:

$$\Rightarrow E_{f,ph} = V_{t,ph} = \frac{11000}{\sqrt{3}} = 6351\ V \tag{8}$$

Normally, the terminal voltage of synchronous generator is assumed as the reference voltage. Therefore:

$$\mathbf{V_{t,ph}} = \mathbf{E_{f,ph}} = 6351\angle 0°\ V \tag{9}$$

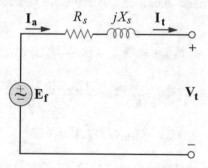

Fig. 6.5. The circuit of solution of Problem 6.7

6.8. Based on the information given in the problem, we have:

$$\mathbf{V_{t,ph}} = \mathbf{E_{f,ph,1}} = 6351\angle 0°\ V \tag{1}$$

$$\mathbf{E_{f,ph,2}} = 1.5\mathbf{E_{f,ph,1}} = 1.5 \times 6351\angle 0°\ V = 9526.5\angle 0°\ V \tag{2}$$

Figure 6.6 shows the per-phase equivalent circuit of a synchronous generator. By applying KVL in the loop for the second condition, we have:

$$\mathbf{E_{f,ph,2}} = \mathbf{V_{t,ph}} + (R_s + jX_s)\mathbf{I_{t,2}} \tag{3}$$

$$\Rightarrow \mathbf{I_{t,2}} = \frac{\mathbf{E_{f,ph,2}} - \mathbf{V_{t,ph}}}{R_s + jX_s} = \frac{9526.5\angle 0° - 6351\angle 0°}{j10} \tag{4}$$

$$\Rightarrow \mathbf{I_{a,2}} = \mathbf{I_{t,2}} = 317.55\angle 90\,^\circ A \tag{5}$$

$$\Rightarrow PF = \cos(\theta_V - \theta_I) = \cos(0 - 90\,^\circ) \Rightarrow PF = 0$$

Since $(\theta_V - \theta_I) < 0$, the power factor is leading.

Choice (1) is the answer.

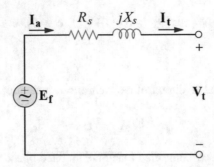

Fig. 6.6. The circuit of solution of Problem 6.8

6.9. Based on the information given in the problem, we have:

$$\mathbf{V_{t,ph}} = \mathbf{E_{f,ph,1}} = 6351\angle 0\,^\circ V \tag{1}$$

$$\mathbf{E_{f,ph,2}} = 0.5\mathbf{E_{f,ph,1}} = 0.5 \times 6351\angle 0\,^\circ V = 3175.5\angle 0\,^\circ V \tag{2}$$

Figure 6.7 shows the per-phase equivalent circuit of a synchronous generator. By applying KVL in the loop for the second condition, we have:

$$\mathbf{E_{f,ph,2}} = \mathbf{V_{t,ph}} + (R_s + jX_s)\mathbf{I_{t,2}} \tag{3}$$

$$\Rightarrow \mathbf{I_{t,2}} = \frac{\mathbf{E_{f,ph,2}} - \mathbf{V_{t,ph}}}{R_s + jX_s} = \frac{3175.5\angle 0\,^\circ - 6351\angle 0\,^\circ}{j10} \tag{4}$$

$$\Rightarrow \mathbf{I_{a,2}} = \mathbf{I_{t,2}} = 317.55\angle - 90\,^\circ A \tag{5}$$

$$\Rightarrow PF = \cos(\theta_V - \theta_I) = \cos(0 - (-90\,^\circ)) \Rightarrow PF = 0$$

Since $(\theta_V - \theta_I) > 0$, the power factor is lagging.

Choice (4) is the answer.

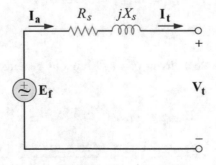

Fig. 6.7. The circuit of solution of Problem 6.9

6.10. Based on the information given in the problem, we have:

$$S = 50 \, MVA \tag{1}$$

$$V_t = 30 \, kV \tag{2}$$

$$f_s = 60 \, Hz \tag{3}$$

$$R_s \approx 0 \, \Omega/phase \tag{4}$$

$$X_s = 9 \, \Omega/phase \tag{5}$$

$$PF = 0.8 \, \text{Lagging} \tag{6}$$

Since the synchronous generator is a star-connected machine, its rated phase voltage is as follows:

$$V_{t,ph} = \frac{30 \, kV}{\sqrt{3}} = 17.32 \, kV \tag{7}$$

Normally, the terminal voltage of synchronous generator is assumed as the reference voltage. Therefore:

$$\mathbf{V_{t,ph}} = 17.32 \angle 0° \, kV \tag{8}$$

The rated current of machine can be calculated as follows:

$$I_t = \frac{S}{\sqrt{3}V_t} = \frac{50 \, MVA}{\sqrt{3} \times 30 \, kV} = 962.25 \, A \tag{9}$$

Solving (6) and (9):

$$\mathbf{I_t} = I_t \angle - \cos^{-1}PF \tag{10}$$

$$\Rightarrow \mathbf{I_t} = 962.25 \angle - \cos^{-1}0.8 = 962.25 \angle - 36.87° \, A \tag{11}$$

In (11), negative sign is applied since the power factor of load is lagging.

Figure 6.8 illustrates the per-phase equivalent circuit of a synchronous generator. By applying KVL in the loop, we have:

$$\mathbf{E_{f,ph}} = \mathbf{V_{t,ph}} + (R_s + jX_s)\mathbf{I_t} \tag{12}$$

$$\Rightarrow \mathbf{E_{f,ph}} = 17320 \angle 0° + (0 + j9)(962.25 \angle - 36.87°) = 23558 \angle 17.1° \, kV \tag{13}$$

$$\Rightarrow \begin{cases} E_{f,ph} = 23558 \, V \\ \delta = 17.1° \end{cases}$$

Choice (1) is the answer.

The value of phase angle of excitation voltage (power angle) is positive because the machine is operated as a generator.

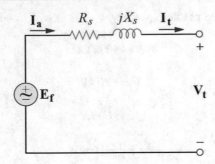

Fig. 6.8. The circuit of solution of Problem 6.10

6.11. Based on the information given in the problem, we have:

$$V_{t,ph} = 17.32\ kV \tag{1}$$

$$E_{f,ph} = 23.558\ kV \tag{2}$$

$$R_s \approx 0\ \Omega/phase \tag{3}$$

$$X_s = 9\ \Omega/phase \tag{4}$$

The theoretical steady-state (static) stability limit of a three-phase generator or the theoretical maximum deliverable power of a three-phase generator can be calculated as follows:

$$P_{max} = \frac{3V_{t,ph}E_{f,ph}}{X_s} \tag{5}$$

$$\Rightarrow P_{max} = \frac{3 \times 23.558 \times 17.32}{9} \Rightarrow P_{max} = 136\ MW \tag{6}$$

A synchronous generator has its theoretical steady-state (static) stability limit if its power angle is 90°. In other words:

$$\delta_{max} = 90° \tag{7}$$

Figure 6.9 shows the per-phase equivalent circuit of a synchronous generator. By applying KVL in the loop, we have:

$$\mathbf{E_{f,ph}} = \mathbf{V_{t,ph}} + (R_s + jX_s)\mathbf{I_t} \tag{8}$$

$$\Rightarrow \mathbf{I_t} = \frac{\mathbf{E_{f,ph}} - \mathbf{V_{t,ph}}}{R_s + jX_s} = \frac{23558\angle 90° - 17320\angle 0°}{j9} \tag{9}$$

$$\Rightarrow \mathbf{I_a} = \mathbf{I_t} = 3248.85\angle 36.32°\ A$$

Choice (3) is the answer.

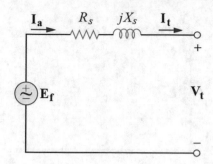

Fig. 6.9. The circuit of solution of Problem 6.11

6.12. Based on the information given in the problem, we have:

$$E_{f,ph,2} = E_{f,ph,1} = 23.558 \, kV \tag{1}$$

$$P_2 = 25 \, MW \tag{2}$$

$$V_{t,ph} = 17.32 \, kV \tag{3}$$

$$R_s \approx 0 \, \Omega/phase \tag{4}$$

$$X_s = 9 \, \Omega/phase \tag{5}$$

The power delivered by the generator in the new condition can be calculated as follows:

$$P_2 = \frac{3V_{t,ph}E_{f,ph,2}}{X_s} \sin \delta_2 \tag{6}$$

$$\Rightarrow 25 \, MW = \frac{3 \times (23.558 \, kV) \times (17.32 \, kV)}{9} \sin \delta_2 \tag{7}$$

$$\Rightarrow \sin \delta_2 = 0.183 \Rightarrow \delta_2 = 10.59° \tag{8}$$

$$\Rightarrow \mathbf{E_{f,ph,2}} = 23.558 \angle 10.59° \, kV \tag{9}$$

Figure 6.10 shows the per-phase equivalent circuit of a synchronous generator. By applying KVL in the loop, we have:

$$\mathbf{E_{f,ph,2}} = \mathbf{V_{t,ph}} + (R_s + jX_s)\mathbf{I_{t,2}} \tag{10}$$

$$\Rightarrow \mathbf{I_{t,2}} = \frac{\mathbf{E_{f,ph,2}} - \mathbf{V_{t,ph}}}{R_s + jX_s} = \frac{23558\angle 10.59° - 17320\angle 0°}{j9} \tag{11}$$

$$\Rightarrow \mathbf{I_{a,2}} = \mathbf{I_{t,2}} = 807.4 \angle - 53.43° \, A \tag{12}$$

$$\Rightarrow PF = \cos(\theta_V - \theta_I) = \cos(0 - (-53.43°)) \Rightarrow PF = 0.596$$

Since $(\theta_V - \theta_I) > 0$, the power factor is lagging.

Choice (1) is the answer.

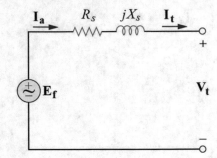

Fig. 6.10. The circuit of solution of Problem 6.12

Index

© The Editor(s) (if applicable) and The Author(s), under exclusive license to Springer Nature Switzerland AG 2023
M. Rahmani-Andebili, *AC Electric Machines*, https://doi.org/10.1007/978-3-031-15139-2

Printed in the United States
by Baker & Taylor Publisher Services